Contents

	Revise		Practise		Review	

KU-540-424

Geographical Skills

You must be able to:

- Demonstrate a range of geographical skills including cartographic, graphical, numerical and statistical.

Cartographic Skills

Latitude and Longitude

- All points on Earth can be located precisely with coordinates using **latitude** and **longitude**.
- Latitude lines provide a measure of how far places are north or south of the Equator. Latitude lines are all parallel to the Equator.
- Longitude lines provide a measure of how far places are east or west from the Prime Meridian (sometimes called the Greenwich Meridian). Longitude lines run between the North Pole and South Pole (meeting at the Poles) at right angles to the latitude lines.

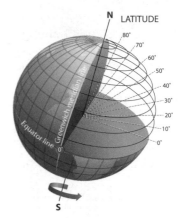

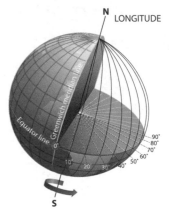

Atlas Maps

- Categories of map include:
  - General: physical (showing **relief**, drainage); human (showing population distribution, population movement, transport networks, settlement layout); **political** (showing borders).
  - Thematic: weather; climate; ecosystems; demographic (population); development; economic (industry).
  - Environmental issues: **climate change**; pollution; desertification; deforestation.

Patterns and Distributions

- You will need to be able to recognise:
  - Patterns: where are things concentrated in **dense** clusters or more **sparse**?
  - Patterns found in: land use; populations; settlements; communication networks (transport and telecommunications); earthquakes; tropical storms.
  - Patterns that have recognisable shapes, e.g. **nucleated**, **linear**, **dispersed** or **radial**.
  - Distributions: describe where things are in a broader sense, using compass points and terms such as central, coastal and **peripheral**.

Ordnance Survey (OS) Maps

- You should be able to locate places using grid lines: give the number of the horizontal **easting line** first, then the vertical **northing line**. A rule may be useful, such as 'along the corridor and up the stairs'.

Collins

GCSE Revision

Geography

Geography

GCSE

Revision Guide

...endan Conway, Janet
...jor and Robert Morris

C000020079905

Contents

		Revise		Practise		Review

- Use four-figure grid references to locate whole grid squares – this is best for larger areas.
- Use six-figure grid references to locate specific points.

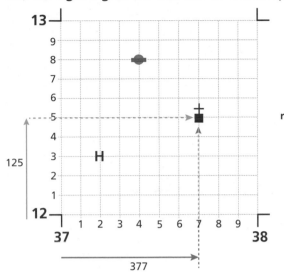

The six-figure grid reference for the church is:

377125

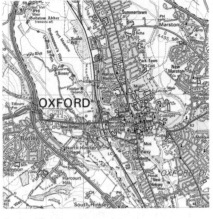

Ordnance Survey mapping

Scale

- All maps use **scale** to provide a standardised way of representing the real world on a page or device screen. Map scales often use a **:** (colon) format, which means 'represents' or 'equal to in reality'. Common OS map scales are 1:25000 and 1:50000.
- Use the scale to check your technique for measuring straight- and curved-line distances on maps.

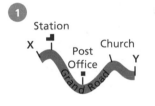

Compass Directions

- Learn the points of the eight-point compass and practise using them to describe locations, e.g. 'the camera is pointing south-west'; 'the river flows in an easterly direction'.

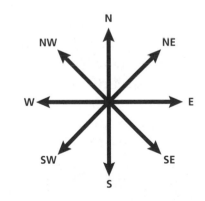

Isolines as Used on Relief and Meteorological Maps

- Relief is the shape or topography of the land, shown by **isolines** called **contours** or by layer colouring.
- Height is usually measured in metres above sea level.
- Points of exact height are shown using **spot heights** or **triangulation pillars**.
- Contours are lines along which the height is the same.
- Steep slopes are shown by closely spaced contours; gentle slopes are shown by contours with larger spaces between them.
- Landscape shapes include **valleys**, **ridges**, **spurs** and **plateaus**.
- Descriptions can also refer to their shape, size, elevation and compass direction (**orientation**).
- Cross-sections are like slices through a landscape, drawn using contours. These can be drawn by hand or with **GIS** (**geographical information system**) elevation profiles.

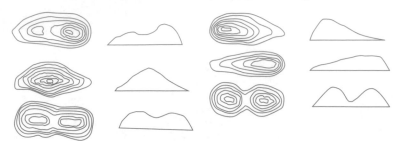

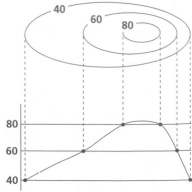

Imagine walking along a straight line from the west side of the hill to the east. You can draw a transect line to show your route. On the transect, put a small dot each time a contour is crossed and label it with its height. Use the line to plot the heights on graph paper or a chart. Then 'join the dots' to show the relief of the hill.

- Helpful terms to describe hills or valleys include **symmetrical** (similar gradient on both sides) or **asymmetrical** (steeper on one side than the other).
- Relief has a significant impact on human activity, e.g. land use and communications.
- Isolines are often used to show weather data such as pressure (**isobars**) and rainfall (**isohyets**).
- They can also be an effective way to map data such as pedestrian counts or pollutants.

Map Symbols

- Always refer to the symbology of maps, using the key provided.
- Practise interpreting by making links between symbols or their absence. For example, on an OS map, a lack of human activity on land beside a river (where there are large gaps between contours) could be due to a desire to avoid flooding.

Drainage: Water on the Land

- River flows or an absence of rivers tell us about the drainage of water from a landscape.
- Drainage density is measured by the total length of river channels in an area, e.g. km per km^2, and is usually high in areas with **impermeable** rocks and low in areas with **permeable** rocks.
- On some rivers channels, **hard** or **soft engineering** is used as a form of **flood management**.
- Drainage includes lakes, waterfalls, underground rivers (e.g. in limestone areas) and artificial water features such as reservoirs.
- Terms used to describe drainage patterns include dendritic, rectangular, parallel, trellised, deranged and radial:

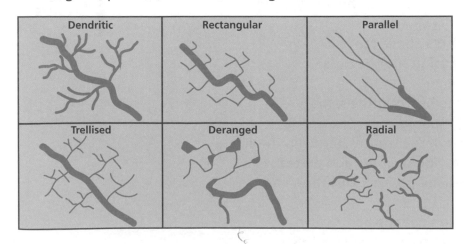

> ### Key Point
>
> You will use your geographical skills when answering questions about all the topics you have studied, and there are lots of opportunities to practise these skills throughout this book.

> ### Key Point
>
> Fundamental geographical skills include the ability to use and understand maps, graphs and charts, and being able to employ numerical data and statistical techniques effectively.

Graphical Skills

- You should be able to suggest appropriate forms of graphical representation. You should also be able to interpret and extract information from, and add to, the following:
  - scatter graphs; line charts; pie charts; bar charts; **histograms** with equal class intervals; particular types of charts such as population pyramids, divided and cumulative bar and line charts
  - pictograms, proportional symbols, **choropleth maps**, isolines, dot maps, desire lines and flow-lines
  - aerial imagery; satellite imagery; false colour images.

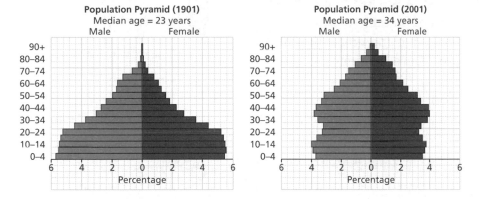

- Fieldwork can often be enhanced by photographs and field sketches, which are really 'diagrams' of a place. Good annotations can add value to graphics.
- Sketch maps are roughly drawn maps that show basic details of an area, and can be labelled or annotated to identify key features.

Numerical Skills

- Number, area and scales; **quantitative** relationships between units.
- Data collection: accuracy; sample size and procedures; control groups; reliability.
- Proportion and ratio; magnitude; **frequency**.

Statistical Skills

- Measures of central tendency: **median**; **mean**; **mode** and **modal class**.
- Measures of spread and cumulative frequency: range; **quartiles** and interquartile range; dispersion graphs.
- Percentages: percentage increase or decrease; **percentiles**.
- Relationships in bivariate data: trend lines through scatter plots; lines of best fit; positive and negative correlation; strength of correlation.
- Predictions and trends: **interpolation**; **extrapolation**.
- Limitations and weaknesses in selective statistical presentation of data.
- Geospatial (geolocated or georeferenced) data presented in a GIS framework. GIS can also be used to analyse spatial data.

Key Words
latitude
longitude
relief map
political map
climate change
dense
sparse
nucleated
linear
dispersed
radial
peripheral
easting line
northing line
scale
isolines
contours
spot heights
triangulation pillars
valley
ridge
spur
plateau
orientation
GIS
elevation profile
symmetrical
asymmetrical
isobars
isohyets
impermeable
permeable
hard engineering
soft engineering
flood management
histogram
choropleth map
quantitative
frequency
median
mean
mode
modal class
quartiles
percentiles
interpolation
extrapolation

Tectonic Hazards 1

You must be able to:

- Understand that natural hazards are the result of physical processes
- Understand how tectonic hazards pose a risk to people and property
- Identify ways that effects from tectonic hazards can be reduced.

Natural Hazards and Their Impacts

- A natural hazard is an extreme event that causes loss of life, severe damage to property or severe disruption to human activities.
- Natural hazards can have a social, economic and environmental impact on an area, and they are more severe in lower income countries (LICs) because there is:
  - more poor quality housing and poor healthcare
  - poor infrastructure so it is harder to reach affected people
  - less money to protect people (e.g. with earthquake-proof buildings) and less money for responses (e.g. providing food).

Earthquakes and Volcanoes

- The Earth's crust is divided into seven large plates and 12 smaller plates that sit on the mantle below. They move very slowly in different directions due to heat convection currents in the mantle:
  - Magma below the surface is heated by the core and rises to the crust, where it cools and sinks, to be reheated once more.
  - Friction caused by the rising currents means that the plates are pushed forward or dragged back (plate tectonics).
  - There are two types of crustal plate: oceanic plates (tend to be denser but thin) and continental plates (less dense and thicker).
  - Two (or more) plates meet at a plate boundary or **plate margin**.
- There are four main types of plate boundary:
  - **Destructive margins** (or convergent boundaries) usually involve an oceanic plate and a continental plate moving towards each other and colliding. The denser oceanic plate is forced beneath the less dense continental plate (a process known as **subduction**) and land is destroyed. The melting plate causes volcanic activity and the stress created by friction between the colliding plates can cause earthquakes.
  - **Constructive margins** (or divergent boundaries) are where plates pull apart, allowing magma to reach the surface and erupt through fissures and faults, creating volcanoes. Earthquake activity also occurs at constructive boundaries.
  - **Conservative margins** are where plates may move in opposite directions to each other, or in the same direction, but at different speeds. As the plates move, their edges stick together, causing friction and stress to build up until one plate jolts forward. This sudden release of energy can cause earthquakes and rift valleys. There is no volcanic activity at conservative margins.

Key Point

The Earth is made up of:

- inner **core**: solid iron and nickel, temperature around 5500°C
- outer core: liquid iron and nickel
- **mantle**: semi-molten rock (magma)
- **crust**: thin layer of solid rock.

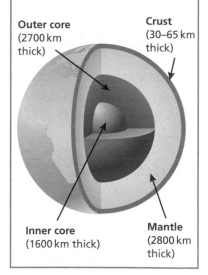

Outer core (2700 km thick)

Crust (30–65 km thick)

Inner core (1600 km thick)

Mantle (2800 km thick)

Key Point

Different physical processes occur at different types of plate boundary.

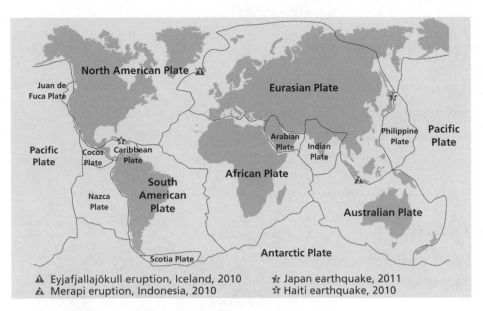

⚠ Eyjafjallajökull eruption, Iceland, 2010
⚠ Merapi eruption, Indonesia, 2010
☆ Japan earthquake, 2011
☆ Haiti earthquake, 2010

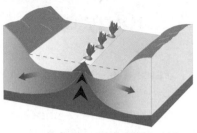

Destructive, e.g. the Nazca Plate subducting beneath the South American Plate

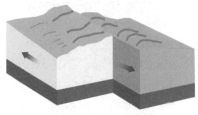

Constructive, e.g. Mid-Atlantic Ridge

- **Collision zones**: if continental plates of the same density collide, they buckle up rather than subduct. This creates fold mountains. There is no volcanic activity at collision zones but earthquakes can occur, e.g. the 2015 Nepal earthquake.
- Most of the world's earthquake and volcano activity is found at the plate boundaries. '**Hot spots**' are exceptions and can occur at any location where the mantle is very close to the surface.
- Volcanic activity is common where the plates meet around the rim of the Pacific Ocean; the area is known as the 'Pacific Ring of Fire'.

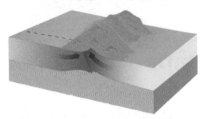

Conservative, e.g. San Andreas fault

Why People Still Live in High-Risk Areas
- They are near friends and family and have a job there.
- Confidence that their government has the ability to 'fix' things.
- Attitude of 'it won't happen to me – it's safe'.
- Volcanic areas have fertile soils, valuable minerals, geothermal energy and tourism opportunities.

Collision zone, e.g. Himalayas and Alps

Reducing the Risk and Effects
- The 3 Ps for locations close to earthquake and volcano zones are:
  - **Plan/predict**: It is possible to predict where earthquakes will occur, but not when. It is possible to predict volcanic eruptions; tell-tale signs include mini-earthquakes, escaping gas and a change in the volcano's shape. Governments can plan evacuations when volcanoes show signs of eruption.
  - **Protect**: Buildings and bridges can be designed to withstand earthquakes and strengthened to withstand the weight of volcanic ash. Firebreaks can reduce the spread of fire.
  - **Prepare**: Emergency services can be trained and prepared. People can be taught how to react in an earthquake or an evacuation. Countries can receive emergency aid or longer-term aid to help rebuild infrastructure and buildings.

Key Point
Monitoring, prediction and planning can reduce the risk from tectonic hazards.

Key Words
core
mantle
crust
plate margin
destructive margin
constructive margin
conservative margin
collision zone
hot spot

Quick Test
1. Describe destructive and constructive plate boundaries.
2. Give two reasons why people still live near active volcanoes.

Tectonic Hazards 2

You must be able to:

- Understand how the effects of and responses to earthquakes vary between areas of contrasting levels of wealth.

Anatomy of an Earthquake

- Focus – point below surface where an earthquake occurs.
- **Epicentre** – the location on the surface directly above the focus.
- Seismic waves – shock waves of energy that travel through rock to the surface.
- Seismogram – measures shaking by an earthquake.
- Aftershocks – smaller tremors in the days after an earthquake.
- **Richter scale** – measures energy released in an earthquake.
- **Primary effects** – occur straight away when a tectonic hazard strikes.
- **Secondary effects** – occur later on, bringing more problems to those affected.

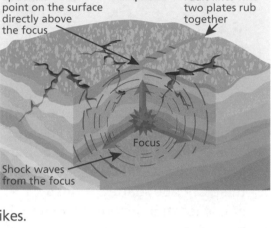

Epicentre – the point on the surface directly above the focus

Earthquakes

Fault lines where two plates rub together

Focus

Shock waves from the focus

Earthquake Case Study – Tohoku, Japan, Higher Income Country (HIC)

Background
- 11 March, 2011 – 2:46 pm.
- Epicentre in the ocean 62 miles (100 km) north-east of Honshu island.
- A complex destructive plate margin involving the Pacific, Okhotsk, Eurasian and Philippine plates.
- 9.0 on the Richter scale.
- The focus was 18 miles (29 km) below the surface.
- Fourth largest earthquake in the world.

Primary Effects
- Buildings destroyed and whole towns demolished.
- Massive damage in Sendai City (1 million people).
- Homes destroyed.
- Loss of power supplies.
- Roads blocked.
- Water shortages.
- Effects felt in Tokyo 185 miles (298 km) away.
- Japan's main island (Honshu) moved 7.8 feet (2.4 metres).

Secondary Effects
- Resulting **tsunami** (with waves up to 40 m high) killed 15 000 people.
- 2 million people left homeless.
- Layers of mud and debris left on land.
- Buildings, vehicles and bodies washed out to sea.
- Workers unable to shut down the Fukushima nuclear reactor, allowing radiation to leak.
- Japan's stock market collapsed.

Aftermath of the 2011 Japan earthquake

Immediate Responses
- Aid came from many other nations.
- A nuclear exclusion zone was established.
- Tsunami warning issued 3 minutes after the earthquake.
- Japanese Red Cross mobilised emergency teams.
- Warning given to other locations, such as Hawaii.

Long-Term Responses
- New, efficient tsunami warning system.
- Modifications to tsunami walls and floodgates that had not been effective.

Earthquake Case Study – Haiti, Lower Income Country (LIC)

Background
- 12 January, 2010 – 4:53 pm.
- Epicentre 16 miles (25 km) south-west of Port-au-Prince city.
- 7.0 on the Richter scale.
- Focus only 10–15 km below surface.
- Haiti is the poorest country in the Western Hemisphere.
- North American Plate slid past the Caribbean Plate.

Primary Effects
are the most dangerous
- 316 000 killed.
- 1 million homeless.
- Difficult to get aid into the country.
- Homelessness, loss of power and roads blocked.
- Airport, port and hospitals closed.

Secondary Effects
– the least dangerous
- 1.6 million people in refugee camps.
- Water and food shortages.
- Increase in crime, particularly looting.
- Outbreaks of cholera.

Immediate Responses
- Aid slow to arrive due to the damaged port and airport.
- The USA sent troops to support an aid programme.
- Many left Port-au-Prince city as their homes had been destroyed.

Long-Term Responses
- Dependence on overseas aid.
- $100 million aid from the USA and $330 million from Europe.
- New homes were built to a higher standard.
- Rebuilding of the port.

> **Key Point**
>
> Countries of contrasting levels of wealth show different effects and responses to tectonic hazards.

One of the tent cities in Port-au-Prince following the 2010 earthquake

> **Key Words**
>
> **epicentre**
> **Richter scale**
> **primary effects**
> **secondary effects**
> **HIC**
> **tsunami**
> **LIC**

> **Quick Test**
>
> 1. What is the difference between the primary and secondary effects of a tectonic hazard? ✓ ✓
> 2. What is the difference between the focus and the epicentre of an earthquake? ✓

Tectonic Hazards 3

You must be able to:

- Understand how the effects of and responses to volcanoes vary between areas of contrasting levels of wealth.

Features of a Volcano

- **Composite volcanoes** are made of alternating layers of lava and ash and are the result of multiple eruptions over hundreds of years. They give explosive eruptions of lava and ash. They are more likely to be found along destructive plate boundaries.
- **Shield volcanoes** form from runny magma (does not trap gases):
  - There is no build-up of pressure and eruptions are not explosive.
  - The runny lava flows a long way from the eruption, creating wide volcanoes.
  - They are more likely to be found on constructive plate boundaries, and at hot spots.
- **Pyroclastic flow** – torrent of hot ash, rocks, gases and steam, moving at up to 450 mph (700 km/h).
- **Ash cloud** – blocks out the Sun.
- **Lahar** – 60 mph mudslide of melted snow and volcanic ash.
- **Active volcano** – erupted recently.
- **Dormant volcano** – currently inactive but might erupt in future.
- **Extinct volcano** – not erupted for many thousands of years.

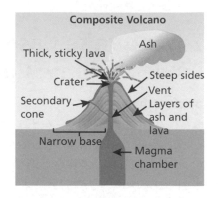

Composite Volcano

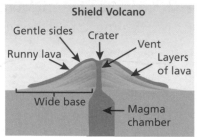

Shield Volcano

Volcano Case Study – Eyjafjallajökull, Iceland, Higher Income Country (HIC)

Background

- Erupted in April, 2010.
- Occurred at the spreading Mid-Atlantic Ridge of the constructive plate margin.
- North American Plate moving west; Eurasian Plate moving east.

Primary Effects

- Volcanic ash plume at 11 000 metres.
- Erupted under the glacier, causing severe flooding.
- Damage to roads, bridges, water supplies and livestock.
- Lava flows.
- Very fine-grained ash was a hazard to air traffic.

Secondary Effects

- Air space closed in Europe and thousands of flights cancelled.
- Glaciers covered in ash for months (increasing glacial melt).

Immediate Responses

- 500 farm families evacuated.
- National Emergency Agency dredged rivers, cleared ash and installed temporary bridges.

The ash cloud from the Eyjafjallajökull eruption caused the cancellation of flights all over the world

Long-Term Responses

- Individual nations grounded air traffic according to local weather conditions following the eruption.
- A scientific review investigated the effect of ash eruptions on air traffic.
- Further research continues to find better ways of monitoring ash concentrations and improving forecast methods.

Volcano Case Study – Mount Merapi, Indonesia, Lower Income Country (LIC)

Background

- Erupted in October, 2010.
- Caused by the Indo-Australian Plate subducting beneath the Eurasian Plate.
- Destructive plate margin in the 'Pacific Ring of Fire'.

Primary Effects

- 353 people were killed.
- 360 000 people were displaced to 700 emergency shelters.
- Volcanic bombs and hot gases for up to 11 km.
- Pyroclastic flows of up to 14 km.
- Ash fell up to 30 km away.
- Villages close to the volcano were buried.

Secondary Effects

- 350 000 people left homes in the area.
- Sulphur dioxide blown across Indonesia and as far as Australia.
- Ash cloud disrupted air transport.
- Roads blocked.
- Lahars.
- Food prices increased.
- Airports closed.

Immediate Responses

- Evacuation centres were established.
- 20 km exclusion zone was set up around the volcano.
- International aid arrived from charities like the Red Cross.

Long-Term Responses

- People moved to newer and safer homes.
- Funds were provided for farmers to replace livestock.
- Data from the eruption contributed to computer models to improve predictions.
- Improved warning and evacuation procedures.
- Construction of dams to hold back future lahars.

> **Key Point**
>
> The effects of and responses to volcanic eruptions vary between areas of contrasting wealth.

The eruption of Mount Merapi sent down tonnes of ash

> **Key Words**
>
> composite volcano
> shield volcano
> pyroclastic flow
> lahar
> active volcano
> dormant volcano
> extinct volcano

Quick Test

1. Which has a wider base: a composite or a shield volcano?
2. What is the difference between pyroclastic flow and a lahar?

Global Circulation System and Atmospheric Hazards

You must be able to:

- Describe and explain global circulation (atmospheric circulation) and its links to atmospheric (hydro-meteorological) hazards
- Describe and explain the structure and features of a tropical storm
- Analyse the possible links between climate change and tropical storms.

The Global Circulation System

- The **global circulation system (atmospheric circulation)** transfers energy via **circulation cells** (e.g. Polar, Ferrel and Hadley cells).
- Alternate low- and high-pressure belts develop, creating a pattern of **climate zones**.
- **Low pressure** is rising air, which creates unstable conditions (stronger winds, clouds, rain).
- **High pressure** is descending air, creating stable conditions (gentler winds, few clouds, little rain).
- At the Equator, low pressure is dominant.
- In the Tropics, high pressure (also called **anticyclones**) is common on land, but above warm oceans, intense low-pressure cells form **tropical storms**.
- In the mid-latitudes, low-pressure systems (called **depressions**) dominate. **Warm** and **cold fronts** occur, producing frontal rain.

> **Key Point**
>
> The air in the atmosphere moves around in circulation cells that transfer and redistribute energy around the Earth.

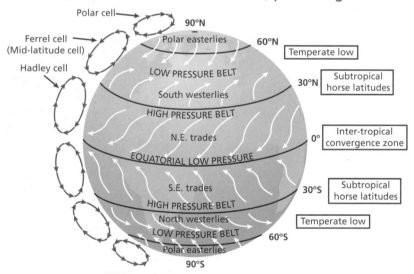

> **Key Point**
>
> Tropical storms are intense low-pressure systems with distinct structure and features formed over warm ocean waters in low latitudes.

- If **blocking high** pressure is dominant for weeks or months, it can cause **droughts**, **heatwaves** in summer and cold frosty/foggy conditions in winter.
- The Earth's rotation creates **Coriolis force**, where winds blow around low and high pressure in opposing directions, which are different in each hemisphere:

Pressure System	N Hemisphere	S Hemisphere
Low	Anti-clockwise	Clockwise
High	Clockwise	Anti-clockwise

Hurricane Katrina, August 2005: a Category 5 tropical storm which devastated New Orleans in southern USA

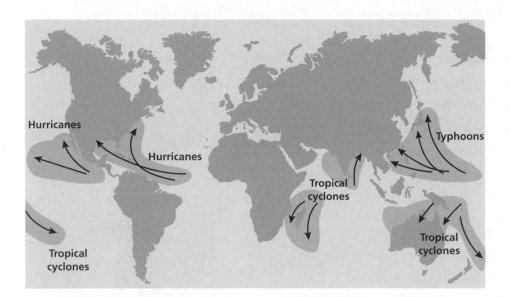

Tropical Storms

- Tropical storms form over warm ocean water ($\geqslant$ 26–27°C) between 5° and 20° north or south of the Equator where the Coriolis force is stronger, which helps initiate the 'spin'.
- Moist air evaporates and rises, creating low pressure. As it cools, condensation occurs, creating clouds and convectional rain.
- A tropical storm can be 1000 km wide, with a distinctive structure.
- If a tropical storm becomes **cyclonic**, air around the centre spins as a vortex creating an eye 30–65 km wide.
- The intense low pressure creates a dome of seawater and a **storm surge** occurs, leading to coastal flooding, which can cause more casualties than the high winds.
- Cooler, drier air is dragged down into the **eye of the storm** – a zone with strangely calm weather and very little cloud or rain. Conversely, around the edge of the eye the winds are strongest.
- The Saffir–Simpson scale measures tropical storm intensity based on wind speed.
- When a tropical storm makes landfall, it loses its energy supply (warm water) and slows due to friction (especially over hilly land).

How Might Climate Change Affect Tropical Storms?

- Warm ocean water is a key driver of tropical storm formation.
- Warmer oceans expand, so storm surges may be worse.
- The distribution of tropical storms and their **frequency** and **intensity** may increase but the evidence for this is inconclusive.

> ### Quick Test
>
> 1. What type of pressure system is a tropical storm? ✓
> 2. What type of weather system can produce heatwaves? ✓
> 3. What causes winds to rotate around weather systems in different directions?
> 4. What feature may be found at the centre of a tropical storm?
> 5. Explain why a tropical storm weakens on landfall.

> ### Key Point
>
> If the oceans become warmer due to climate change, it may affect the distribution, frequency and intensity of tropical storms.

Structure of a Tropical Storm

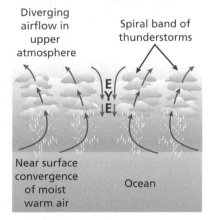

> ### Key Words
>
> global circulation system
> atmospheric circulation
> circulation cells
> climate zones
> low pressure
> high pressure
> anticyclones
> tropical storm
> depression
> warm front
> cold front
> blocking high
> drought
> heatwave
> **Coriolis force**
> cyclone/cyclonic
> storm surge
> eye of the storm
> frequency
> intensity

Tropical Storms – Case Study

You must be able to:

- Demonstrate that you can analyse a case study of a tropical storm
- Describe and explain the primary and secondary effects
- Describe the immediate and long-term responses
- Evaluate some of the responses and explain how the 3 Ps can reduce the negative effects.

LIC Case Study: Typhoon Haiyan

Where, When and Why did it Happen?

- The Philippines (capital: Manila), South-East Asia.
- Gross national income per capita: US$7500 (emerging economy).
- November, 2013.
- Causes included warm oceans in the western Pacific and it is believed that climate change may have contributed.

What were the Main Effects and Impacts?

- A Category 5 tropical storm.
- The strongest-ever storm recorded at landfall.
- Wind speeds up to 315 km/h (195 mph).
- **Primary effects** included:
  - widespread devastation of property, especially in Leyte
  - landslides
  - storm surges: a storm shelter was submerged and many drowned.
- **Secondary effects** included:
  - **infrastructure** damage which impeded relief efforts
  - economic impacts such as businesses closed and development halted
  - social impacts such as homelessness (1.9 million), disease, and schools closed
  - environmental impacts to ecosystems and farmland damaged
  - human factors which made matters worse (**3 Ps**):
    prediction – high-level warnings were too late; **protection** – communications infrastructure was too vulnerable; **planning** – too many people stayed at home in spite of warnings (**inertia**).

Who was Affected and How did People Respond?

- The death toll was over 6300 (mostly due to the storm surge).
- Immediate responses included:
  - much of the Philippines was put under a state emergency
  - worldwide relief effort totalled over $500 million.
- Long-term responses (how did or can management strategies reduce risk?):
  - governance heavily criticised
  - 3 Ps reviewed: authorities adopted proactive strategies
  - resilience was improved by tackling inertia; for example, incentives were successfully offered to encourage evacuation and when Typhoon Hagupit (also Category 5) hit the Philippines in 2014, only 18 people were killed.

The Philippines

South China Sea · Philippine Sea · LUZON · EASTERN VISAYAS · VISAYAS · LEYTE · Tacloban · Sulu Sea · MINDANAO · Malaysia

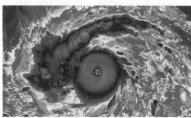

A satellite image of Typhoon Haiyan

> ### Key Point
>
> The effects of a tropical storm on a country depend significantly on its **resilience**, which in turn depends on its level of development and preparedness.

Devastation after Typhoon Haiyan

HIC Case Study: Hurricane Sandy

Where, When and Why did it Happen?

- North-eastern USA (capital: Washington DC).
- Gross national income per capita: US$54 629 (highly developed, tenth richest in world per capita).
- October 22–November 2, 2012.
- Warm surface water in the Caribbean and Atlantic.
- Climate change may have contributed.

What were the Main Effects and Impacts?

- A Category 2–3 tropical storm.
- Largest Atlantic hurricane ever, spanning 1800 km (1100 miles).
- High wind speeds of up to 185 km/h (115 mph).
- **Primary effects** included:
  - wind damage to infrastructure
  - floods due to heavy rain and storm surges
  - subway tunnels were flooded
  - flights cancelled
  - over 6 million customers lost power supplies
  - 7000 people in emergency shelters
  - New York Stock Exchange closed for two days.
- **Secondary effects** included:
  - economic: second costliest storm ever in USA (over $71.4 billion)
  - social: homelessness (650 000)
  - political: may have affected the presidential election
  - military: led to a review of climate change impact on national security
  - environmental: untreated sewage released into sea.

Who was Affected and How did People Respond?

- Death toll: 157 in the US (233 in total).
- Immediate responses included:
  - Federal Emergency Management Agency (FEMA) coordinated the 3 Ps
  - many eastern states were declared states of emergency and widespread evacuations took place
  - public services (e.g. schools) were closed
  - President Obama arranged federal financial assistance.
- Long-term responses (how did or can management strategies reduce risk?):
  - as the richest country in the world, the USA was well-resourced and the key players well-trained in implementing the 3 Ps, creating a high level of resilience
  - lessons had been learned from Hurricane Katrina in 2005.

Then president, Barack Obama, personally supervised preparations for the arrival of Hurricane Sandy and visited the damaged areas afterwards

Quick Test

1. For Typhoon Haiyan and Hurricane Sandy, compare:
   a) The death toll
   b) The magnitude
   c) One primary and one secondary effect
   d) The causes
   e) One immediate and one long-term response

UK Climate and Extreme Weather Events

You must be able to:

- Demonstrate that you can explain the main influences on the UK climate
- Describe the main types of atmospheric hazards affecting the UK
- Analyse the causes, impacts and responses of an extreme weather event in the UK.

How Does the UK's Geographic Location Affect its Climate?

- The UK has a temperate climate, i.e. relatively moderate temperatures and precipitation.
- A key influence is its location in the **mid-latitudes** on the coast of north-west Europe beside the Atlantic, where there is frequent conflict between cold and warm **air masses** which create **fronts** and these produce precipitation.
- The **prevailing winds** (the most common winds) are westerlies from the Atlantic Ocean, and the relatively warm, moist air from the **North Atlantic Drift** (or **Gulf Stream**) creates a dominant **maritime** effect.
- Less frequently, the UK receives relatively dry **continental** air from land masses to the south and east. In winter, such air can be extremely cold but, in summer, very hot.
- **Altitude** is also influential. Higher ground in the west has greater precipitation (**relief rainfall**) and lower temperatures. Consequently, the east is in a **rain shadow**, receiving much lower precipitation.

Key Point

The main atmospheric hazards impacting the UK include winter storms, snow and ice, and drought.

Key Point

Extreme weather events such as drought have economic, social and environmental impacts.

Types of Weather Hazard in the UK

- The main hazards affecting the UK are atmospheric or **hydro-meteorological hazards** linked to the water cycle.
- The main types of weather hazard event include: river flooding; sea flooding; winter storms; snow and ice; drought.

An Extreme Weather Event in the UK: Drought 2010–12

- Droughts are extended dry periods lasting for months or years.
- The 2010–12 drought was one of the ten most significant of the last 100 years in the UK.

Causes

- Blocking highs (slow-moving anticyclones) led to very dry winters in 2009–10 and 2011–12.
- East winds were common, bringing in dry continental air.

A Weather Chart Showing a Blocking High

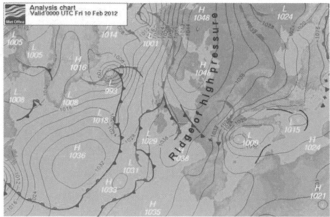

- Significantly lower precipitation than normal (see map, below right).
- Climate change may have contributed to the drought.
- An imbalance between the demand and supply of water in the densely populated south-east and Midlands, which suffer **water deficit** and **water stress**. These areas rely on the piping of water from regions of **water surplus** and/or groundwater supplies.

Impacts

- Economic and social impacts:
  - farmers struggled to provide water for livestock and to harvest crops
  - low reservoir levels.
- Environmental impacts:
  - groundwater and river levels were very low, affecting aquatic ecosystems
  - wildfires spread.

Management Strategies

- Hosepipe bans were introduced (affecting 6 million consumers).
- Water meters were installed to monitor usage.
- Water companies fixed leaking pipes.
- Water-saving devices were encouraged.
- Education about water use was improved.
- Improved wastewater recycling.
- New reservoirs and pipelines were considered.
- Desalination plants were considered (but high cost).

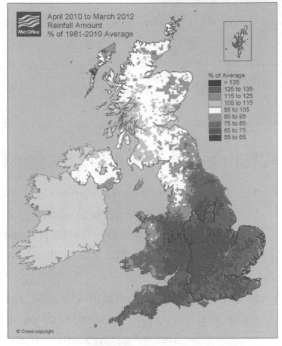

April 2010 to March 2012
Rainfall Amount
% of 1981-2010 Average

% of Average
> 135
125 to 135
115 to 125
105 to 115
95 to 105
85 to 95
75 to 85
65 to 75
55 to 65

© Crown copyright

> **Key Point**
>
> The UK drought of 2010–12 had economic, social and environmental impacts.

Evidence that UK Weather is Becoming More Extreme

- Flood warnings are increasing. A BBC news report on February 9, 2014, based on an interview with Dame Julia Slingo, the Met Office's chief scientist, said there had been 130 severe flood warnings since December, 2013, compared with nine in the whole of 2012.
- Greater frequency of extreme weather, such as intense rainfall in sudden downpours, hot days and named winter storms.
- Vineyards are becoming viable in wider areas of the UK (e.g. Wales, Hampshire, West Sussex).
- **Jet stream** has more variable loops, which are wider/deeper.
- Scientists are now increasingly certain of human impact (**anthropogenic**) on climate change.

> **Quick Test**
>
> 1. What is meant by a maritime influence on the UK climate?
> 2. a) Give an example of one extreme weather event in the UK.
>   b) Describe one impact and one management strategy for it.
> 3. Give one indication that the UK weather is becoming more extreme.

> **Key Words**
>
> **mid-latitudes**
> **air masses**
> **front**
> **prevailing wind**
> **North Atlantic Drift**
> **Gulf Stream**
> **maritime**
> **continental**
> **altitude**
> **relief rainfall**
> **rain shadow**
> **hydro-meteorological hazards**
> **water deficit**
> **water stress**
> **water surplus**
> **jet stream**
> **anthropogenic**

Climate Change 1

You must be able to:

- Describe and explain the main evidence for and causes of climate change
- Assess the range and uncertainty of climate change projections.

What is the Evidence for Climate Change?

- The Intergovernmental Panel on Climate Change (**IPCC**) is an internationally accepted authority on climate change.
- The IPCC found that Northern Hemisphere temperatures during the 20th century were the highest for the past 1300 years.

Long-term Evidence

- In the absence of reliable climate data, **proxy measures** are used, e.g. ice cores, marine sediment cores and pollen analysis.
- Proxy measures of temperature since the start of the **Quaternary period** (last 2.6 million years) show that the Earth has experienced an 'Ice Age' of several lengthy **glacials** (extremely cold periods of glacier growth) interspersed with shorter **interglacials** (warmer periods of glacier retreat).
- The Earth is currently in an interglacial period.

Short-term Evidence (Last Few Hundred Years)

- As climate measurements are unreliable before 1850, other proxy measures are needed such as tree rings and historical sources (e.g. landscape paintings and literature).
- Data show a marked global temperature rise since 1850 (the 'hockey stick' graph): warming oceans; sea level rise; reduction in Arctic sea ice; possible increases in frequency and intensity of extreme weather events.
- The atmosphere creates a natural **greenhouse effect**, allowing short-wave solar radiation in (e.g. light), but **greenhouse gases** prevent the escape into space of some of the heat created (long-wave infrared radiation).
- Industrialisation and population growth have increased greenhouse gas emissions.
- Carbon dioxide increase was first noticed in the 1950s and 60s by scientists in Hawaii, with record highs in 2015.
- Global temperatures in 2015 reached 1°C above the 'pre-industrial' period of 1850–1900.
- The 11 warmest years in the instrumental record have occurred since 1998.

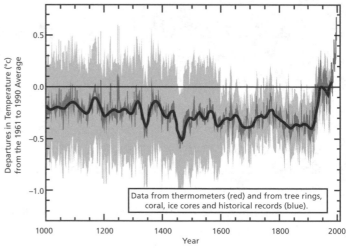

'Hockey Stick' Graph: Variations in Northern Hemisphere Temperature Over the Last 1000 Years

Data from thermometers (red) and from tree rings, coral, ice cores and historical records (blue).

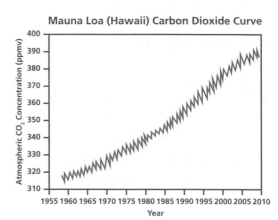

Mauna Loa (Hawaii) Carbon Dioxide Curve

Climate Change Causes

- Natural or physical factors include:
  - orbital changes (Milankovitch cycles)
  - solar output (changes in the Sun's energy)
  - asteroid/comet collisions.
- Human (anthropogenic) factors include:
  - use of fossil fuels producing greenhouse gases, e.g. power generation and transportation
  - agriculture, e.g. methane from cattle and rice paddies
  - deforestation, e.g. tree removal reduces natural **carbon sequestration**
  - methane release from melting permafrost and ocean floors, e.g. owing to anthropogenic global warming.

Revise

Range and Uncertainty of Projections

- The IPCC reviews, researches and makes projections using computerised **climate models** of the atmosphere, creating a **range** of possible outcomes called **scenarios**, depending on levels of greenhouse gas emissions.
- The graph shows how temperatures have increased between 1900 and 2000 by about 0.5°F and how they are predicted to change in the future. The lower emission scenario is based on significant changes in the sources of energy used, with greater consumption of renewable energies and more use of efficient technologies. The even higher emission scenario is based on the continuing consumption of non-renewable carbon-based sources, continued population growth and further industrialisation in the emerging countries.

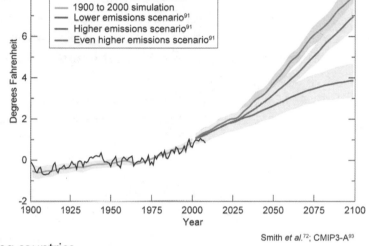

Global Average Temperature 1900–2100

- 1900 to 2008 observations
- 1900 to 2000 simulation
- Lower emissions scenario[91]
- Higher emissions scenario[91]
- Even higher emissions scenario[91]

Smith *et al.*[72]; CMIP3-A[93]

- There is a degree of uncertainty about the projections owing to:
  - unpredictable physical factors, e.g. solar radiation and volcanic eruptions which can lead to atmospheric cooling
  - unpredictable human factors, e.g. it is too soon to assess the impact of target-setting, economic trends and new low-carbon technologies
  - trends are difficult to spot with a relatively short period of accurate observations
  - uncertainty about the influence of the upper atmosphere, sea currents and deep ocean temperatures.

Climate Change 2

You must be able to:

- Assess different approaches to the management of climate change
- Describe and explain the main impacts of climate change
- Describe and explain El Niño and La Niña.

Managing Climate Change Impacts

- The IPCC recommends **adaptation** and **mitigation**.
- Adaptation involves making adjustments to reduce potential damage, e.g. improving flood defences; taking advantage of new opportunities, e.g. some crops can be grown at higher latitudes.
- Mitigation involves cutting greenhouse gas emissions; reducing or eliminating long-term risk to human life and property; internationally agreed targets to reduce greenhouse gas emissions; **carbon capture (carbon sequestration)**.
- Carbon capture removes carbon at emission source and stores it deep underground.
- Critics of adaptation say that although necessary, without mitigation it only treats the symptoms, not the causes of, climate change.
- Critics of mitigation say that reversal of climate change is not possible. A small minority deny it completely.

Consequences of Climate Change

- Consequences of climate change will be global, but variable.
- Most impacts will be harmful, but some could be beneficial.
- **Global weirding**: weather events are likely to become more extreme.
- Ocean acidification owing to absorption of carbon dioxide by seawater, with negative impacts on marine ecosystems (e.g. coral bleaching).
- Increased evaporation is likely to lead to higher precipitation.
- The Arctic is warming faster than equatorial regions: the Arctic Ocean may become ice-free in the summer, producing an **enhanced greenhouse effect**, as there is a lower albedo and so more heat will be absorbed rather than reflected.
- Sea levels are increasing owing to **thermal expansion** and melting of land-based glaciers.
- Low-lying coastal areas are under threat as coastal urban populations are increasing in most countries.
- Soil moisture reduction is affecting farm yields and viability.
- Permafrost melting could release large quantities of methane (a potent greenhouse gas).
- Flora and fauna of biomes are already responding. Some species are adapting and extending the range of their habitats, but others are struggling to survive.

Key Point

Climate change can be managed by adaptation or mitigation; both strategies have their critics.

How Climate Change Could Affect the World

 The average temperature rise across the globe could be 4°C but in the Arctic the increase could be as much as 16°C. Extreme heatwaves could threaten lives, e.g. New York could see summer temperatures over 50°C.

 Sea levels could rise by 80 cm by the end of the century, while storm surges could pose a serious threat to countries like the Netherlands.

 River flow could be reduced in regions such as the Mediterranean and southern Africa, causing water shortages. Droughts could occur twice as frequently in these areas.

 Maize and wheat yields could fall by 40% in Africa and rice yields by 30% in Asia, leading to food shortages and increased risk of starvation in vulnerable countries.

 Oceans are likely to acidify with the absorption of increasing amounts of carbon dioxide, causing serious harm to marine ecosystems.

 Higher temperatures are likely to result in more forest fires in areas such as the Amazon, Australia, southern Europe and the USA.

Causes and Impacts of El Niño/La Niña

- El Niño's proper name is the **El Niño Southern Oscillation (ENSO)**. It has occurred many times over hundreds of years, usually around Christmas (El Niño is Spanish for the 'Christ Child').
- It occurs in the Pacific Ocean at irregular intervals of between two and seven years, and lasts nine months to two years.
- Surface waters in the eastern Pacific, near the west coast of Central and South America, become warmer than normal.
- Consequently, low-pressure systems develop where high pressure is the norm and the region receives well-above average rainfall, causing flooding and landslides.
- The warmer seawater lacks nutrients so the marine food web collapses, reducing fish stocks and damaging the fishing industry.
- On the western side of the Pacific, opposite effects occur: whereas low pressure and high rainfall are the norm, high pressure occurs bringing much less rain and leading to droughts and wildfires.

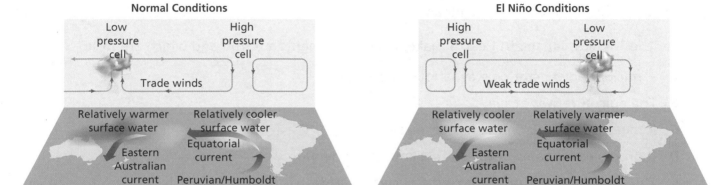

- **La Niña** (Spanish for 'The Girl') refers to conditions which are the opposite of El Niño (colder waters in the eastern Pacific; warmer in the western Pacific). The impacts are usually less severe.
- The impacts of El Niño and La Niña are thought to extend well beyond the Pacific. Such long-distance effects are called '**teleconnections**'. The 2015–16 El Niño was the strongest since 1950 and is thought to have had global impacts on weather events.
- Climate change may increase the frequency and intensity of El Niño and La Niña events.

Quick Test

1. Which of these climate change responses are **not** adaptation? Which are **not** mitigation?
   - a) electric cars
   - b) higher sea walls
   - c) tidal power
   - d) wind farms
   - e) IPCC carbon reduction targets
   - f) improving air conditioning in houses
2. Where does seawater become warmer in an El Niño episode?

Tectonic Hazards 1

1 What causes the Earth's plates to move? [2]

2 Draw a diagram to help explain what happens at a destructive plate margin. [4]

3 Draw a diagram to help explain what happens at a constructive plate margin. [4]

4 Draw a diagram to help explain what happens at a conservative plate margin. [4]

Total Marks .... / 14

Tectonic Hazards 2

1 What do the terms LIC and HIC mean? [2]

2 For the Tohoku, Japan 2011 earthquake, which statements are true and which are false?

A A nuclear power station was seriously damaged.

B The epicentre was at sea.

C It measured over 6 on the Richter scale.

D It produced tsunami waves up to 40 metres in height.

E The tsunami waves only affected a nuclear reactor. [5]

3 Explain why the largest earthquakes do not always cause the most deaths. [6]

Total Marks .... / 13

Tectonic Hazards 3

1 Identify two primary effects following the eruption of Eyjafjallajökull. [2]

2 What long-term responses can be adopted to reduce the effects of a volcanic eruption? [4]

Total Marks .... / 6

Global Circulation System and Atmospheric Hazards

1 Which of the following statements is true?

A Tropical storms form along the Equator.

B Tropical storms form between latitudes 5° and 20° north and south of the Equator.

C Tropical storms form between latitudes 30° and 50° north and south of the Equator.

D Tropical storms form between latitudes 40° and 60° north and south of the Equator. [1]

2   Give three key features of a tropical storm such as the one shown right.   [3]

3   What are the regionally specific names used for tropical storms in these areas?

    a)   Indian Ocean and near Australia

    b)   Atlantic, Caribbean and Pacific (near North and South America)

    c)   Pacific (near Asia)   [3]

4   How does the Coriolis force (Coriolis effect) affect wind direction around tropical storms in different parts of the world?   [4]

> **Total Marks** .... / 11

Tropical Storms – Case Study

1   What are the 3 Ps?   [3]

2   For a tropical storm you have studied, state its name, where and when it occurred.   [3]

3   Choose the main category of these impacts of a tropical storm: [5]

Impacts	Economic	Social/Political	Environmental
Homelessness			
Factories and other businesses closed or inaccessible due to damage to transport infrastructure			
Waterborne diseases			
Damage to ecosystems			
Schools closed for weeks			

4   For these effects of a tropical storm, indicate which are primary (P) and which are secondary (S).

    a)   People drowned in storm surge.

    b)   Thousands of people are homeless for over a year.

    c)   Outbreaks of cholera and dysentery kill hundreds of people.

    d)   Severe damage to infrastructure such as railways and bridges by flash floods.

    e)   Homes destroyed by high winds.

    f)   Schools and businesses are closed for months.   [6]

> **Total Marks** .... / 17

Practice Questions

UK Climate and Extreme Weather Events

1 For an extreme weather event in the UK that you have studied, state when it occurred. [1]

2 How does altitude of the land influence the UK climate? [2]

3 What are the main types of weather hazard affecting the UK? [3]

4 Outline the main causes of a named extreme weather event in the UK. [4]

Total Marks .... / 10

Climate Change 1

1 IPCC is the abbreviated name of which organisation?

   A International Policy on Climate Change

   B International Panel on Climate Change

   C Intergovernmental Policy on Climate Change

   D Intergovernmental Panel on Climate Change

   E Intermediate Panel on Climate Change [1]

2 What term is sometimes used to describe the human causes of climate change? [1]

3 Give three human causes of climate change. [3]

4 Draw lines to match the key terms to the correct definitions. [3]

Key term	Definition
Glacial	Indirect ways to find out average temperatures from the past
Interglacial	Last 2.6 million years, including the 'Ice Age'
Quaternary	Warmer periods of glacier retreat
Proxy measure	Extremely cold periods of glacier growth

Total Marks .... / 8

Climate Change 2

1 Which of the following is the correct definition of mitigation?

   A Making adjustments to reduce potential damage; limiting the impacts; taking advantage of new opportunities.

   B Cutting greenhouse gas emissions; reducing or eliminating long-term risk to human life and property; internationally agreed targets to reduce greenhouse gas emissions. [1]

2 Explain why in the eastern Pacific (e.g. off the coast of Peru), El Niño events can cause problems for the local fishing industry. [2]

3 Sort the following statements about the consequences of an El Niño event into the correct parts of the table. [2]

- Low-pressure systems develop where high pressure is the norm.
- Surface waters cooler than normal.
- Surface waters warmer than normal.
- Below average rainfall, causing droughts and wildfires.
- Well above average rainfall, causing flooding and landslides.
- High-pressure systems develop where low pressure is the norm.

	Western Pacific	Eastern Pacific
Surface waters		
Pressure		
Rain		

4 For these management of climate change strategies, decide which is adaptation (A) and which is mitigation (M).

   a) Developing more drought-resistant crops or irrigation schemes

   b) Making buildings more energy efficient

   c) Higher flood defences along coasts and rivers

   d) Greater use of renewable resources

   e) Carbon capture and storage

   f) Planting more trees [6]

Total Marks .... / 11

Ecosystems and Balance

You must be able to:

- Describe how ecosystems are structured using key words
- Explain, using a small-scale UK example, how ecosystems can have their balance disrupted.

What is an Ecosystem?

- **Ecosystems** are communities of **biotic** (living) and **abiotic** (non-living) components that all function together to create a distinctive environment.
- Ecosystems can be small scale, such as a pond or hedgerow in the UK.
- Ecosystems can be large scale, such as a **tropical** rainforest, and these are known as 'biomes'.
- Ecosystems are balanced – each organism plays its part and can be easily disrupted, particularly by the actions of humans.
- **Food chains** show simple relationships between different organisms. Basically that means, what eats what.
- **Food webs** show more complex **interrelationships** and **interdependence** between organisms.
- **Producers** are organisms such as plants that convert the Sun's energy into sugars thus producing food for them.
- **Consumers** are species that eat other species. They can be herbivores (such as goats) that eat plants, or carnivores (such as lions) that eat other animals.
- **Decomposers** such as fungi act to break down the remains of dead plants and animals and then return their nutrients to the ecosystem.
- This transference of nutrients is called the **nutrient cycle**.
- These diagrams show a food chain and a food web.

A pond is a small-scale ecosystem

> ### Key Point
>
> Ecosystems are a critical part of the natural world. All species alive today form part of an ecosystem. The climate and soil form the building blocks of all ecosystems.

Food Chain of an Owl

Producer	Herbivore/ consumer	Omnivore/ consumer	Carnivore/ consumer

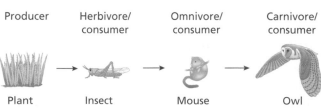

| Plant | Insect | Mouse | Owl |

A Food Web

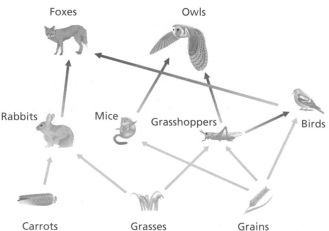

Foxes · Owls · Rabbits · Mice · Grasshoppers · Birds · Carrots · Grasses · Grains

Balance in Ecosystems

- Every part of the Earth's surface is part of an ecosystem. Important examples include hot deserts, tropical rainforests, tropical grasslands, **temperate** deciduous forests, temperate grasslands, **boreal** forests, **polar** regions, tundra and coral reefs.
- Maintaining balanced ecosystems is critical to the survival of life on Earth.
- A small-scale example of an ecosystem in the UK is a hedgerow. In a hedgerow, the plants – such as small trees – are the producers, small animals such as mice are herbivore consumers and birds such as owls are carnivore consumers.
- Ecosystems naturally manage themselves in a variety of complex ways, such as populations of particular organisms being kept in check by being the prey of another. Too many beetles born in one year means more food for the birds that prey on them.

A hedgerow is another type of small-scale ecosystem

Human Use of Resources

- Human activity can disrupt the balance of an ecosystem. An example of this is deforestation as the removal of trees destroys the habitats of countless organisms, thus leading to potential imbalance. The logical endpoint is the disappearance of a particular ecosystem.
- The biosphere – the surface and atmosphere of the Earth that is occupied by living organisms – provides humans with simple resources such as food, water and building materials.
- Other more complex resources that humans can extract from the biosphere include medicines such as aspirin and quinine, poisons for use in hunting, and fuel.
- Humans have exploited the biosphere in an unsustainable way.
- Animals such as the great auk, a flightless bird, have been hunted to extinction.
- Certain trees, such as mahogany and rosewood, have been cut down in huge numbers for their beautiful timber and others have been exploited for fuelwood.
- The large-scale burning of **fossil fuels**, such as coal, oil, peat and natural gas, has led to the release of huge amounts of carbon dioxide into the atmosphere, which has in turn driven global climate change. Coal and oil release sulphur dioxide when they burn, causing breathing problems and contributing to acid rain.

Key Point

An ecosystem functions properly when all of its parts work in harmony. This harmony can be disrupted by human actions.

Key Words

ecosystem
biotic
abiotic
tropical
food chain
food web
interrelationship
interdependent
producer
consumer
decomposer
nutrient cycle
temperate
boreal
polar
fossil fuel

Quick Test

1. Give an example of a large-scale and a small-scale ecosystem.
2. What is the difference between a food web and a food chain?
3. What do the terms 'biotic' and 'abiotic' mean?
4. Demonstrate how nutrients move around an ecosystem.
5. Name two harmful gases released by the burning of fossil fuels.

Ecosystems and Global Atmospheric Circulation

You must be able to:

- Describe how global atmospheric pressure changes between January and July
- Describe the locations of a range of global ecosystems.

Global Atmospheric Circulation

- Differences in **atmospheric pressure** exist. Atmospheric pressure fluctuates constantly but patterns of air movement can be observed.
- Pressure patterns can be broadly predicted and follow seasonal trends.
- Wind is the movement of air owing to many factors, such as the movement of air masses.
- Air masses can be disrupted and their behaviour distorted by features such as mountains, coasts and rivers.
- **Wind patterns** can be predictable but may be extremely erratic.
- The locations of the world's major ecosystems are heavily influenced by the existence and movement of the Earth's major high- and low-pressure areas and the resulting wind patterns.

> **Key Point**
>
> Atmospheric circulation affects where ecosystems are found.

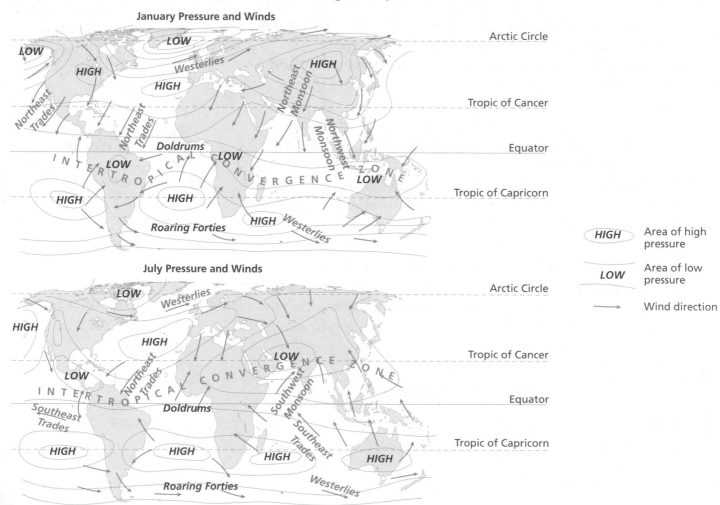

Global Ecosystems

Tropical rainforests
- Found in a wide belt encircling the Earth, roughly following the Equator and for the most part between the Tropics of Cancer and Capricorn.
- The Amazon, in the northern part of the Southern Hemisphere, is the world's largest rainforest.
- The climate is hot and wet all year round.

Tropical grasslands
- Named after the Tropics of Cancer and Capricorn, along both of which they can be found.
- The most famous tropical grasslands are those of the savannas of central Africa.
- They have long dry and brief wet seasons.
- They have little tree cover.

Temperate grasslands
- Found in two wide belts to the north of the Tropic of Cancer and to the south of the Tropic of Capricorn.
- They flourish in the centres of the continents of North America (prairies) and Asia (steppe).
- They have warm summers with very cold winters.
- They have little tree cover.

Temperate forests
- Found mostly in the Northern Hemisphere, well to the north of the Tropic of Cancer but to the south of the Arctic Circle.
- Examples of note are the deciduous forests of north-western Europe and eastern China.
- They have warm, damp summers and mild winters.
- They have deciduous trees.

Boreal forests
- Found in a thin belt just to the south of the Arctic Circle, fringing tundra-type ecosystems.
- Examples of places with boreal forests include central Canada.
- They have mild summers and very cold winters.
- The trees are mostly coniferous.

Tundra
- Ecosystems found in a wide belt encircling the Earth in what are known as the high latitudes – areas on and a little south of the Arctic Circle.
- Much of northern Russia, Canada and Iceland have tundra-type environments.
- They are very cold regions.
- They have little tree cover.

Polar regions
- Found in the far north, above the Arctic Circle, and far south, below the Antarctic Circle.
- Greenland in the north and Antarctica in the south are examples of polar regions.
- The climate is very cold all year round, but with little precipitation.

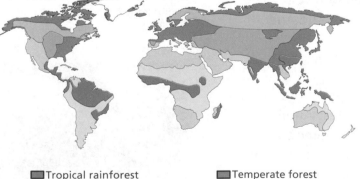

■ Tropical rainforest
■ Tropical grassland
■ Hot desert (see pages 36–37)
■ Temperate grassland
■ Temperate forest
■ Boreal forest
■ Tundra
□ Polar regions

Quick Test
1. Where are tropical rainforests found?
2. What things can affect the behaviour of air masses?
3. Give examples of places that have temperate forests.

 Key Words

atmospheric pressure
wind patterns

Ecosystems in the UK

You must be able to:
- Describe the distribution of UK ecosystems
- Explain how humans exploit the biosphere as a resource.

Ecosystems in the UK

- The UK's main land ecosystems are moorlands, heaths, woodlands and wetlands.
- The UK's natural ecosystems thrive even as the human population grows, and are diverse.

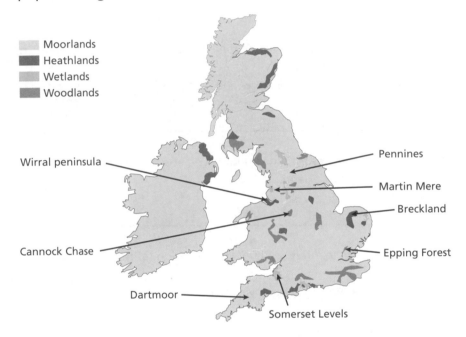

Moorlands
Heathlands
Wetlands
Woodlands

Wirral peninsula
Pennines
Martin Mere
Breckland
Cannock Chase
Epping Forest
Dartmoor
Somerset Levels

Moorlands

- Upland moorland covers large swathes of the UK, including the Pennines of Yorkshire and Dartmoor in Devon.
- **Biodiversity** is limited owing to the high **altitude** and cold, windy and wet conditions. The characteristic plant is heather.

Combestone Tor near Yelverton on Dartmoor

Heathlands

- Heathlands are typical of lower-lying areas, such as on the Wirral peninsula in Merseyside and Breckland in East Anglia.
- Heathlands exhibit moderate biodiversity and are home to many species of trees, such as birch, and plants such as heather and gorse. Reptiles such as sand lizards are found here.
- Owing to rich soil and flat **relief**, heathlands have been popular for farming and **settlement**. Estimates suggest that only 6% of the UK's natural heathland remains.

Heathland merges into a landscape of trees towards the town of Glossop, near Manchester

Wetlands

- Wetlands are coastal- or river-based ecosystems formed between the land and the water. They are found in river **estuaries** or **floodplains**. UK examples include Martin Mere in west Lancashire and the Somerset Levels.
- Wetlands have moderate biodiversity including water- and salt-tolerant plants such as rushes.
- They provide an important habitat for many species of bird, fish and other water-loving animals such as otters.
- Many of the UK's wetland areas have been drained for farming, settlement or industry and the rest are threatened by these factors.

Key Point

Wetland wildlife is threatened by the drainage and pollution of these ecosystems.

Restored wetlands, Somerset Levels

Woodlands

- There is much woodland in the UK but only about 20% of this can be described as 'ancient' – in existence since before 1600 in England.
- Woodlands have a moderate biodiversity and a high level of interdependence or harmony among living and non-living elements. Plants include bluebells, oaks and birches.
- Much of the UK's woodland is managed through controlled access and protection of Sites of Special Scientific Interest (SSSIs).
- Examples of woodland in the UK include Epping Forest, on the north-east edge of London, and Cannock Chase, in Staffordshire.

Autumn woodland in Epping Forest, London

Key Point

The UK has a diverse landscape and group of ecosystems.

Quick Test

1. Why does the moorland ecosystem exhibit only limited biodiversity?
2. What type of ecosystem can be found on the Wirral in Merseyside?
3. Where is Martin Mere?
4. Why have many of the UK's wetlands been drained?

Key Words

biodiversity
altitude
relief
settlement
estuary
floodplain

Tropical Rainforests

You must be able to:

- Describe what makes a tropical rainforest so distinctive
- Describe opportunities, threats and management strategies in tropical rainforests such as the Amazon.

Characteristics of Tropical Rainforests

- Tropical rainforests are located in the tropics, a band around the Equator from 23.5° north to 23.5° south.
- There are biotic and abiotic elements to tropical rainforests.
- Nutrients are rapidly recycled in tropical rainforests by decomposers such as fungi.
- Trees play a critical part in the water cycle in rainforests as they discharge huge amounts of water through **transpiration**.
- There can be a huge variety of tree species in a rainforest (as many as 300 per square kilometre).
- The climate in rainforests is generally hot all year round, with average daily temperatures of 27–29°C. Rainfall is also high at around 500–600 mm each month.
- The soil is known as a **latosol**. Latosols are deep but have few nutrients. The majority of the nutrients are found in the **leaf litter**, which decays rapidly owing to the warm, damp climate.
- Rainforests develop in layers – we call this **stratification**:
  - above the forest floor is a shrub layer of small bushes surviving in the dark conditions
  - the under canopy of young trees receives limited sunlight
  - the canopy has most life and the upper parts of most trees
  - the emergents are the tops of the tallest trees and get most sun.
- Due to the huge amounts of light, heat and water, rainforest ecosystems have high biodiversity.

Adaptations in the Rainforest

- Trees such as the mahogany in the Amazon have developed huge **buttress roots** that provide stability and absorb nutrients directly from the soil.
- Pitcher plants have adapted to the low nutrient soils by developing a taste for insects – they are attracted using scent glands and then caught using a slippery flower before being digested.
- Forest elephants eat clay from ponds in clearings to counteract the toxins in the leaves they eat.
- Chimpanzees are omnivores – they will consume vegetation or other animals in order to broaden their diet.
- The rainforest is an example of **interdependence** – each species plays its part and can be easily disrupted, especially by humans.

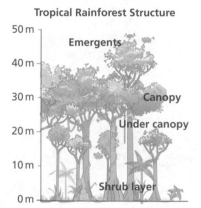

Tropical Rainforest Structure

Emergents
Canopy
Under canopy
Shrub layer

50 m
40 m
30 m
20 m
10 m
0 m

Key Point

The soils, plants and animals in a tropical rainforest have all developed to work together in harmony – they are interdependent.

Buttress roots

Opportunities, Threats and Management Strategies in the Amazon

- Tropical rainforests provide goods and services to humans:
  - **Hunter-gatherers** collect plants and catch wild animals to eat
  - Soils can be fertilised through small-scale **shifting cultivation** of crops such as manioc and cassava
  - Trees provide fuelwood and building materials
  - Medicines and hunting poisons can be extracted from a variety of plants and animals.
- **Commercial** and **subsistence farming** of crops (such as soybeans) and animals (such as cows) are major reasons for the deforestation of the Amazon.
- There are large deposits of minerals such as iron, bauxite, nickel and copper beneath the forest, but to reach them means large-scale removal of the vegetation.
- Building roads such as the Trans-Amazonian Highway has led to increased settlement as Brazil's population has grown. Roads also allow access to harder-to-reach areas, thus creating new centres of expansion.
- **Hydroelectric power (HEP)** schemes, such as that at Tucuruí in Brazil, have created huge amounts of energy for a growing economy but have also flooded huge areas of forest.
- Removing the vegetation creates soil erosion as roots that bind the soil together no longer do so.
- With deforestation comes a loss of habitats and a resultant drop in biodiversity as species die out.
- The fastest way to clear forest is through burning it, but this has led to huge amounts of carbon dioxide being released.
- Brazil has experienced positive **economic development** as dams have provided energy for industry. Mining and farming have provided jobs to millions and given Brazil a source of export income.
- Rainforests can be sustainably managed by:
  - selective logging (cutting), whereby only commercially valuable trees above a certain height are felled, thus leaving smaller trees to attain maturity
  - international agreements that make illegal the export of endangered plant and animal species
  - encouraging **ecotourism**, which makes the forest itself the tourist attraction
  - using labelling schemes such as the international Forest Stewardship Council (FSC), to encourage the purchase of trees grown in sustainable conditions.

Key Point

Despite international outcry, the Amazon Rainforest in South America continues to be deforested at an alarming rate.

Ecotourism can make an attraction of rainforests

Quick Test

1. Describe the climate in tropical rainforest.
2. Name two rainforest plants and their adaptations.
3. Give three reasons for deforestation.

Key Words

transpiration
latosol
leaf litter
stratification
buttress roots
hunter-gatherers
shifting cultivation
commercial farming
subsistence farming
hydroelectric power (HEP)
economic development
ecotourism

Hot Deserts

You must be able to:

- Describe the locations and distinctive characteristics of hot deserts
- Describe the challenges and opportunities for a named hot desert such as the Mojave in south-western USA.

Characteristics of Hot Deserts

- Hot deserts are found in two broad belts encircling the Earth, between 15 and 30° north and south of the Equator.

Global Distribution of Hot Deserts

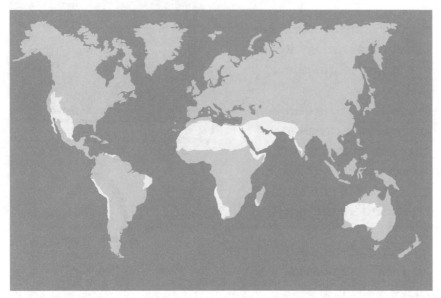

- Temperatures in hot deserts are very high at 30–45°C. Rainfall is extremely low at around 3 mm each month. Water is scarce.
- Rivers appear and disappear quickly, often as the result of high rainfall in mountain sources.
- Water in some deserts appears as snow, which some animals like the Bactrian camel have adapted to by eating the snow.
- The soil is poor and supports little vegetation.

> **Key Point**
>
> Biodiversity in hot deserts is much lower than in other ecosystems but the species that live there are adapted to both the climate and other species.

Adaptations in Hot Deserts

- The saguaro cactus of the Mojave Desert in North America has no leaves in order to cut down moisture loss through transpiration. It also has shallow but wide-ranging roots so that it can take advantage of brief rainstorms.
- The quiver tree of southern Africa has fleshy leaves that allow it to retain moisture.
- Fennec foxes are nocturnal, which allows them to avoid the **extreme temperatures** of the day. The fennec also has large ears in relation to the rest of its small body and they act as radiators, letting heat escape.

Fennec foxes

Opportunities, Threats and Management Strategies in the Mojave, South-Western USA

- In richer countries, governments build **infrastructure** such as roads and bridges.
- In the deserts of south-western USA, large-scale settlement has been made possible by the development of huge water projects such as the Hoover Dam on the Colorado River.
- Extensive commercial farming of crops, such as peanuts, has been made possible by the building of the dam.
- The dam both controlled the Colorado River and created Lake Mead, which today is the source of water used for **irrigation**.
- The Hoover Dam also exists as a source of hydroelectric power (HEP).
- Tourism, such as to the casinos and the other attractions of Las Vegas, has become the biggest industry in the Mojave as a result of the benefits from the construction of the Hoover Dam.
- Increased **desertification** is a real problem in places like the Mojave and other semi-arid areas.
- Increased water use by farming, tourism and settlement means less for the environment.
- Desertification has been accelerated across the globe by things like climate change, population growth, overgrazing by farmed cattle and over-cultivation.
- Strategies to deal with desertification include:
  - encouraging local people to use less water
  - developing planning laws that restrict the size of buildings
  - planting trees to stabilise sand dunes and to also provide shade, fuelwood and fodder for animals
  - encouraging the use of drip irrigation.
- Humans develop the poor soils in hot deserts through irrigation or by using organic mulch to protect it.
- However, over-irrigation can cause **salinisation**, which occurs when the water in soils evaporates in high temperatures, drawing salts to the surface which are toxic to many plants.

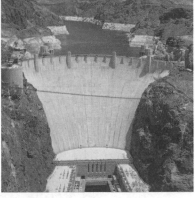

Hoover Dam

Key Point

By using dams to overcome the challenges of the environment, many new challenges are created.

Key Words

extreme temperatures
infrastructure
irrigation
desertification
salinisation

Quick Test

1. Describe the climate in hot deserts.
2. Name two hot desert plants and their adaptations.
3. Give three reasons why desertification has increased.
4. How can humans deal with the threat of desertification?

Coral Reefs and Deciduous Woodlands

You must be able to:

- Describe the location of named coral reefs around the world, nutrient cycling within them, their value to humans, threats to them and management of those threats
- Describe the biotic and abiotic characteristics of and interdependence in temperate deciduous woodlands, their value to humans, threats to them and management of those threats.

Characteristics of Coral Reefs

- Coral reef ecosystems are found in warm tropical seas such as those around Australia.

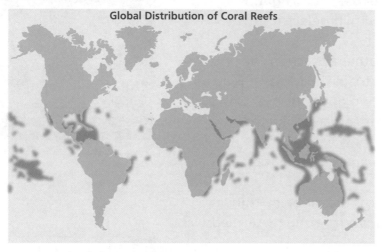

Global Distribution of Coral Reefs

- Coral reefs are biodiverse because they contain huge amounts of nutrients washed into the sea by rivers. The **nutrient cycle** in coral ecosystems focuses on the movement of nitrates.
- When fish die, they decompose at the bottom of the sea (or are eaten by another predator that then excretes their remains onto the sea floor). This organic material is then re-absorbed by bacteria and algae, which return the nitrates to the system.
- The coral, fish and climate are all interdependent.
- Humans are also interdependent with reefs as they provide a source of fish, but the reefs are easily destroyed by trawlers.
- Reefs **sequester** huge amounts of carbon as they act like giant forests absorbing carbon dioxide in their growth process.
- Climate change in the form of **global warming** is affecting the ability of coral to absorb carbon dioxide.
- Reefs can be damaged by pollution flowing from estuaries.
- Humans can cause lasting harm to reefs by breaking pieces off.
- Coral reefs can be sustainably managed through:
  - ecotourism such as that available on Australia's Heron Island, which seeks to use the reef as the attraction itself
  - controlling amounts of industrial waste and sewage released into marine ecosystems
  - action on global warming through controls on fossil fuels
  - setting up marine parks to control tourist numbers.

Key Point

The world's largest coral reef is the Great Barrier Reef, which is found off the north-eastern coast of Queensland, Australia.

Ocean fish and coral reef

Key Point

Coral reefs have very high biodiversity and provide a vital habitat for a huge variety of species.

Characteristics of Deciduous Woodlands

- **Deciduous** woodlands are found 40 to 60° north and south of the Equator. They are found in locations such as north-western Europe and north-eastern USA.

Global Distribution of Deciduous Woodlands

- The climate in deciduous woodlands is temperate.
- The winter season sees the highest rainfall, with places in the USA receiving over 100 mm per month.
- Daily temperatures are between 5 and 7°C in December and January and between 15 and 25°C in June and July.
- Trees have broad leaves that lose water through high rates of transpiration.
- Trees adapt to the cooler winters by shedding their leaves and developing a thick layer of bark to protect themselves against frosts.
- Deciduous woodlands have a relatively rich biodiversity.
- Animals gather food in the summer to eat in the winter during hibernation.
- Many species migrate to warmer regions to avoid the winter cold.
- The soils in deciduous forests are rich, encouraging a wide array of plant life. The rich soil creates an ideal environment for the farming of crops such as wheat.
- Humans use deciduous woodlands for recreation.
- The timber industry is a large provider of jobs and profits to many countries.
- Less than 25% of the world's ancient deciduous forests are intact.
- Global warming promotes the migration of new species into stable forest ecosystems.

Deciduous woodland in South Bavaria, Germany

Key Point

Deforestation of deciduous woodland has led to the need for sustainable management.

Quick Test

1. Name three places around the globe where large coral reefs can be found.
2. How do coral reefs sequester carbon?
3. What are the characteristics of deciduous trees?
4. How do humans use deciduous forests?

Key Words

carbon sequestration
global warming
deciduous

Polar and Tundra Environments

You must be able to:

- Describe the physical characteristics of – and the adaptation of – species to polar and tundra environments and development opportunities therein
- Describe the challenges of developing a cold environment such as Antarctica and strategies to protect it, while balancing the needs of development.

Physical Characteristics of Polar and Tundra Environments

- Polar and **tundra** environments are both found in what are known as the high latitudes. Tundra is found about 60–70° north.

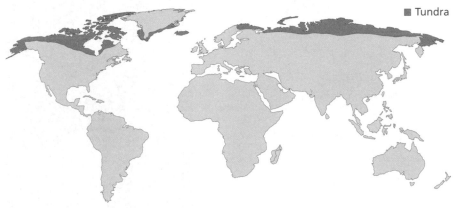

■ Tundra

- Climates in polar and tundra environments vary throughout the year, with places like Alaska and Norway receiving moderate summers that can present daytime temperatures of 20–25°C. Winter temperatures can plunge below –40°C.
- The characteristic landscape in tundra regions is that of **permafrost**.

> ### Key Point
>
> Polar and tundra are both regarded as extreme environments. Both are cold and have characteristic ecosystems and climates. There are development opportunities present in both of these cold environments.

Adaptations

- Soil in tundra regions may remain frozen all year round. Frost-resistant plants such as the Arctic poppy survive by developing adaptations such as shallow roots and flowers that track the path of the Sun.
- Biodiversity in tundra regions is very low due to extreme cold.
- Permafrost, soil, plants and animals are all interdependent and at risk of change due to human actions.
- Animals such as the Arctic fox develop thick coats to protect against the cold.
- The Arctic hare has small ears to reduce heat loss and white fur to avoid the gaze of predators.
- There are huge deposits of minerals such as bauxite, tar sand and standard crude oil.
- Less profitable sources of oil, such as tar sand, have begun to be developed.

Arctic poppies

- The extraction of tar sand oil creates huge numbers of jobs but it destroys habitats.
- Polar ocean regions have rich seas, brimming with fish, and this attracts large-scale industrial fishing. An example is the Bering Sea near Alaska.

Antarctica – Development and Conservation

- Antarctica is the Earth's most southerly continent. There are no permanent settlements, although scientists from many countries occupy temporary stations in various places.
- Antarctica is classed as a desert due to its spectacularly low level of precipitation.
- Antarctica has huge mineral deposits such as coal, bauxite and crude oil.
- Antarctica is highly inaccessible as there is no infrastructure, no means of growing crops to sustain a population, and any buildings would have to be constructed to withstand the extreme climate.
- Commercial fishing is big business in Antarctica, with crews from around the world (but especially Argentina and Chile) making large profits on huge catches. Governments must consider fishing quotas in order to balance the needs of fishermen and the protection of biodiversity.
- **Extreme tourism** is a recent development because tourists are being increasingly attracted to new and often hostile places.
- International pressure groups, such as Greenpeace, campaign for the protection of Antarctica.
- The 1959 Antarctic Treaty is an agreement by 12 countries declaring Antarctica a scientific preserve.
- The Madrid Protocol prohibits all mining in Antarctica to preserve it as a **wilderness** area.

Tar sand

> ### Key Point
>
> Antarctica is the world's last true wilderness and, as such, is in need of protection. Development opportunities exist but must be carefully weighed against the need to conserve the fragile habitats and ecosystems found within Antarctica.

Quick Test

1. Describe the climate of polar and tundra regions.
2. What is permafrost?
3. What are the benefits and drawbacks of extracting oil from tar sands?
4. What is extreme tourism?

> ### Key Words
>
> tundra
> permafrost
> extreme tourism
> wilderness

Review Questions

Tectonic Hazards 1

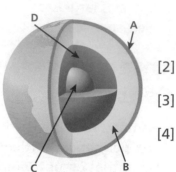

1 What is a 'hot spot'? [2]

2 State three reasons why tectonic hazards have a greater impact in LICs. [3]

3 Identify the different layers of the Earth labelled in this diagram. [4]

4 Describe how the '3 Ps' (predict, protect and prepare) can help to reduce the effects of an earthquake. [6]

> Total Marks _____ / 15

Tectonic Hazards 2

1 What name is given to the standard scale used to measure energy release in an earthquake? [1]

2 In what ways may the effects of an earthquake in a higher income country be different to those experienced in a lower income country? [6]

> Total Marks _____ / 7

Tectonic Hazards 3

1 Draw a diagram to help describe and explain the shape of a shield volcano. [4]

2 Match up the following key terms with the correct definitions.

Key term	Definition
Ash cloud	Torrent of hot ash, rock, and gases and steam
Lahar	Blocks out the Sun, causing suffocation and health problems
Pyroclastic flow	Volcano nobody expects to erupt ever again
Extinct volcano	Volcano that has erupted in past 2000 years but is not currently active
Dormant volcano	Mudslide including rock debris and water

[4]

> Total Marks _____ / 8

Global Circulation System and Atmospheric Hazards

1 What is transferred around the planet by the global circulation system of winds? [1]

2 Explain why coastal flooding may be caused by tropical storms. [3]

3 Explain how climate change may affect tropical storms. [3]

4 Complete the table below, using these terms to fill the gaps:

Cyclones   Typhoons

Mexico   Australia

Country	Name used for Tropical Storms
	Hurricanes
Philippines	
	Cyclones
Bangladesh	

[2]

> Total Marks _____ / 9

Tropical Storms – Case Study

1 Which of the following is most likely to have caused Hurricane Sandy?

A   High pressure in the western Pacific

B   Warm surface water in the western Pacific

C   Cold surface water in the western Atlantic

D   Warm surface water in the western Atlantic

E   High pressure in the western Atlantic [1]

2 In the context of hazard management, what is 'inertia'? [2]

3 How can human factors contribute to a higher death toll when tropical storms hit poorer countries such as the Philippines? [3]

4 Use the data in the table below to compare the impacts of two tropical storms. Comment on the economic impact and death toll, suggesting reasons for the differences.

Name of Tropical Storm (Year)	Country and Continent	Intensity (Saffir–Simpson Scale)	Economic Cost (US$ converted to 2015 values)	Deaths	Population in Area Affected
Cyclone Sidr (2007)	Bangladesh, Asia	5	1.95 billion	4234	146.5 million
Hurricane Ike (2008)	Florida, USA, North America	5	25.4 billion	5	18.4 million

[4]

> Total Marks _____ / 10

Review Questions

UK Climate and Extreme Weather Events

1 Describe two of the main impacts of the UK drought of 2010–12. [2]

2 Give three strategies that were used to manage the UK drought of 2010–12. [3]

3 Explain how prevailing wind influences the UK climate. [3]

4 Draw arrows on the map to match the data below with each of the four climate zones of the UK.

Cold winters, cool summers

Mild winters, cool summers

Cold winters, warm summers

Mild winters, warm summers

Place A	Annual rainfall 3000 mm	January average temperature 8°C	July average temperature 19°C
Place B	Annual rainfall 1000 mm	January average temperature 9°C	July average temperature 20°C
Place C	Annual rainfall 600 mm	January average temperature 3°C	July average temperature 15°C
Place D	Annual rainfall 660 mm	January average temperature 7°C	July average temperature 21°C

[3]

Total Marks _____ / 11

Climate Change 1

1 Which key term is used to describe IPCC projections?

A Scheme

B Schema

C Scene

D Sequence

E Scenario [1]

2 What name is given to the changes in the Earth's orbit that can cause climate change? [1]

3 How does the Earth's atmosphere create a natural greenhouse effect? [2]

4 Give two important reasons for uncertainty about climate model projections. [2]

Total Marks _____ / 6

Climate Change 2

1 ENSO is the abbreviated name of which climate event?

   A El Niño Southern Ocean

   B El Niño Southern Oscillation

   C El Niña Southern Ocean

   D El Niña Southern Oscillation

   E La Niña Southern Oscillation [1]

2 Why is global warming likely to produce higher precipitation? [2]

3 Explain why coral bleaching is likely to occur in marine ecosystems. [3]

4 Choose the correct word in each of these sentences about the consequences of climate change.

- Consequences will be **local / global**.
- Most impacts are likely to be **harmful / beneficial**.
- Weather-related hazards will become **more / less** extreme in intensity.
- Climate change will cause ocean **alkalinisation / acidification**.
- Sea-level change due to thermal **contraction / expansion** of oceans.
- Permafrost melting could lead to **methane / carbon dioxide** release. [3]

> **Total Marks** _____ / 9

Practice Questions

Ecosystems and Balance

1. Give an example of one small-scale ecosystem. [1]

2. Define the term 'ecosystem'. [2]

3. What is the difference between a food chain and a food web? [2]

4. Describe how the nutrient cycle works. [4]

> Total Marks .... / 9

Ecosystems and Global Atmospheric Circulation

1. In January, is high or low pressure found over central Asia? [1]

2. Name two features that can distort the behaviour of an air mass. [2]

3. Using examples, describe the global distribution of tropical rainforests. [4]

> Total Marks .... / 7

Ecosystems in the UK

1. Name one place in the UK that has a heathland ecosystem. [1]

2. What is the characteristic plant of the moorland ecosystem? [1]

3. What word describes the range of species in a given area? [1]

4. Name the UK's four main land-based ecosystems. [3]

> Total Marks .... / 6

Tropical Rainforests

1. What is selective logging? [2]

2. What is the difference between commercial and subsistence farming? [2]

3 Describe stratification in a rainforest.   [4]

4 Describe how tropical rainforests provide goods and services to humans.   [4]

Total Marks .... / 12

Hot Deserts

1 What word describes roads, bridges, water and power lines?   [1]

2 What do the letters HEP stand for?   [1]

3 Using examples, describe the global distribution of hot deserts.   [3]

4 Describe how the problem of desertification can be tackled in the Mojave Desert.   [4]

Total Marks .... / 9

Coral Reefs and Deciduous Woodlands

1 What word describes the way in which species in an ecosystem all rely on each other?   [1]

2 Name the world's largest coral reef and where it can be found.   [2]

3 Describe the movement of nitrates around the nutrient cycle in a coral reef.   [3]

4 Describe the climate of a deciduous forest.   [3]

Total Marks .... / 9

Polar and Tundra Environments

1 Where are polar and tundra environments found?   [1]

2 Name one benefit and one drawback of tar sand oil extraction.   [2]

3 Describe the temperatures over the year in polar and tundra environments.   [2]

4 Using examples, describe how plants have adapted to life in polar and tundra climates.   [3]

Total Marks .... / 8

Glaciation 1: Rocks, Weathering and Mass Movement

You must be able to:

- Identify the three rock types, give named examples and describe their distribution in the UK
- Understand why different rock types are associated with distinctive scenery
- Explain what different weathering processes there are
- Describe mass movement and understand the relationship between mass movement and shape of a slope.

The Three Rock Types

Igneous

- **Igneous** rocks are formed by cooled molten magma (underground, making them intrusive) or by lava (on the surface, making them extrusive). Igneous rock is **impermeable**.
- Granite is an example of an intrusive igneous rock. It is speckled pink, grey and black with glassy quartz crystals and is coarse-grained.
- Basalt is an example of an extrusive igneous rock. It is very fine-grained, black or dark grey.

Sedimentary

- **Sedimentary** rocks are made from particles of other rocks that have been broken or dissolved, then resettled or redeposited.
- Most are **permeable** and/or **porous**, e.g. limestone, chalk, sandstone, clay and mudstones.
- Chalk is soft, white to grey and fine-grained.
- Sandstone is yellow or red with sand and glassy quartz grains.

Metamorphic

- **Metamorphic** rocks have been changed by great heat, pressure or chemical action. The depth at which this happens affects the characteristics of the rock. Any type of rock can be altered, and more than once.
- Metamorphic rocks are impermeable and types include slate, schist and gneiss.
- Slate is fine-grained, grey or green. It has changed from mudstone or shale.
- Schist is a medium to coarse-grained layered rock. It has changed from clays, mudstones or shales.

Weathering

- **Weathering** is the breaking up or decomposition of rock without moving it away. It is a response to changing temperature, pressure and moisture.
- **Physical** or **mechanical weathering** results from frost-shattering, salt crystal growth or unloading (pressure release). The rock disintegrates but is not altered.
- **Chemical weathering** results from the work of oxygen, carbon dioxide or plant acids, diluted in water. The rock decays and allows the removal of minerals.

Key Point

Distinctive landscape features for rock types are:

- tors on granite
- dry valleys on chalk
- clints and grykes on limestone.

Distribution of Rock Types

- Igneous rocks
- Metamorphic rocks
- Sedimentary rocks

- Rainwater is weakly acidic and reacts effectively with limestone.
- Organic acids from decayed plants can react with minerals in rocks. The products of reactions are carried away by rainwater.
- Chemical weathering is usually faster at higher temperatures and when there is more water available.
- **Biological weathering** is the disturbance and breaking up of rocks by plant roots, swaying trees and burrowing animals.

Mass Movement

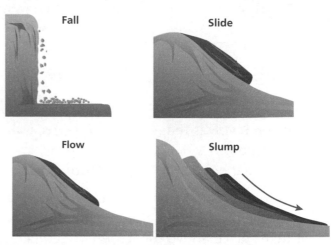

- **Mass movement** of surface material can be very sudden and fast, to imperceptibly slow. The speed is influenced by the surface slope and water content.
- Surface material is easily saturated on impermeable rocks, so is more likely to move.
- Fast mass movement:
  - **Falls**: detached rocks topple from a steep face
  - **Slides**: blocks of material detached from the surface move at the same speed over the underlying rock
  - **Flows**: finer, saturated material moving downslope as a 'tongue' shape; the front and top moves fastest
  - **Slumps**: downward and outward movement of segments of rock along a concave line of weakness.
- Slow mass movement:
  - **Creep**: extremely slow; drier material
  - **Solifluction**: slow; wetter material.

Slope Classification and Development

- Slopes are classified as follows:

≤2°	2–5°	5–10°	10–18°	18–30°	30–45°	>45°
Level	Gentle	Moderate	Moderately steep	Steep	Very steep	Cliff

- **Retreat**: slope wears away backwards but keeps the same shape.
- **Decline**: slope gets lower and gentler.
- **Replacement**: slope shape changes as weathered material stays at the base.

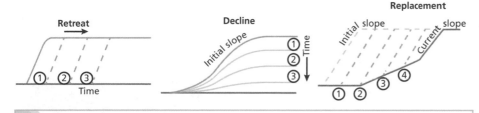

> ### Key Point
>
> There may be long periods of very slow mass movement interrupted by a sudden, fast episode of movement.

Key Words

igneous	fall
impermeable	slide
sedimentary	flow
permeable	slump
porous	creep
metamorphic	solifluction
weathering	retreat
mass	decline
movement	replacement

Quick Test

1. Which rock type has been changed by heat, pressure or chemical action?
2. Name one fast and one slow form of mass movement.
3. Would you expect dry or wet debris on a hillside to move quickly?

Glaciation 2: Ice Processes

You must be able to:

- Understand the connection between weathering and erosion
- Describe and explain the weathering associated with the presence of ice on a landscape
- Understand the difference between glaciers and ice sheets
- Describe and explain the processes of glacial erosion
- Explain why and how moving ice can transport material
- Explain where and why material is deposited by ice.

Weathering and Erosion

- Weathering breaks up rock but does not move it.
- **Erosion** involves the removal of debris, usually weathered, from its location.
- Some further weathering of debris can occur during transport.

Weathering Associated with Low Temperatures

- Freeze-thaw action is also known as **frost-shattering**.
- It takes place above the ice where temperatures go above and below freezing point.
- Water enters cracks or weaknesses in a rock, then expands upon freezing.
- Ice takes up more space than water, so pressure is put on the rock, weakening it.
- Rock is shattered into **angular** fragments called **scree**.
- Land under an **ice sheet** will not be weathered, but will be eroded.

Glacial Erosion

- A **glacier** cannot 'rip' rocks from the landscape.
- To erode effectively, ice needs to have debris in it.
- **Abrasion** is when rock fragments at the base of a glacier scour, scrape and polish the **bedrock** over which the glacier is moving.
- The rock fragments need to be harder than the bedrock.
- Rocks that have been abraded have scratches or **striations** on them.
- **Plucking** is when cracks in the bedrock will collect water melted from a glacier. This water refreezes and expands, causing the crack to widen.
- Over time, the rock will break away and be carried along in the glacier.

> ### Key Point
>
> The thicker the ice, the more erosion is likely to take place.

Abrasion

Particles at the base of the ice scratch the bedrock

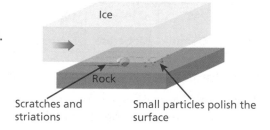

Scratches and striations

Small particles polish the surface

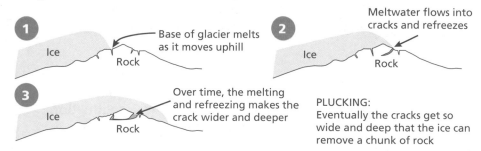

PLUCKING:
Eventually the cracks get so wide and deep that the ice can remove a chunk of rock

Rock surface on Snowdon showing evidence of abrasion and plucking

- A newly glacier-eroded landscape is rough and rocky.

- Ice sheets are huge bodies of ice that completely cover the landscape.
- Glaciers will get trapped into their valleys under an ice sheet and even stop moving.

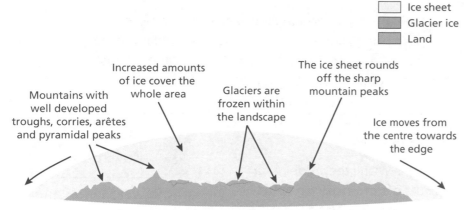

Ice sheet
Glacier ice
Land

Mountains with well developed troughs, corries, arêtes and pyramidal peaks

Increased amounts of ice cover the whole area

Glaciers are frozen within the landscape

The ice sheet rounds off the sharp mountain peaks

Ice moves from the centre towards the edge

The weight of the ice depresses the land

Glacial Transport and Deposition

- Glaciers transport material that has fallen onto the ice from weathered surfaces above and which the moving ice has eroded from its base and sides.
- Debris carried by a glacier is called **moraine**.
- Debris size ranges from tiny particles to huge boulders.
- The location of the debris identifies the moraine:
  - in front of the ice: **terminal moraine**
  - under the ice: **sub-glacial moraine**
  - inside the ice: **en-glacial moraine**
  - on top of the ice: **supra-glacial moraine**
  - at the side of the ice: **lateral moraine**
- Where two glaciers meet, their lateral moraines may form a **medial moraine**.
- Material moves up and down within the ice.
- **Deposition** results from a reduction in the size and energy of a glacier.
- Deposition happens most rapidly at the front and edges of a glacier.
- Deposition is most evident in lowland areas.
- Deposits vary from loose rocks (**erratics**) to blankets of material (**till**).

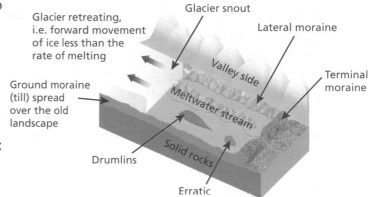

Glacier retreating, i.e. forward movement of ice less than the rate of melting

Glacier snout

Lateral moraine

Ground moraine (till) spread over the old landscape

Valley side

Meltwater stream

Terminal moraine

Drumlins

Solid rocks

Erratic

Quick Test

1. What is material carried by ice called?
2. What shape are frost-shattered rocks?
3. Where is a lateral moraine?
4. What does rock that has been abraded look like?

Key Words

erosion	sub-glacial moraine
frost-shattering	en-glacial moraine
angular	supra-glacial moraine
scree	
ice sheet	
glacier	
abrasion	lateral moraine
striations	
bedrock	medial moraine
plucking	
moraine	deposition
terminal moraine	erratic
	till

Glaciation 3: Glacial, Physical and Human Landscapes and Post-Glacial Change

You must be able to:

- Describe and explain the formation of glacial erosion landforms
- Describe and explain the formation of glacial deposition landforms
- Understand the spatial relationships of features in glaciated landscapes
- Understand how post-glacial conditions affect the landscape
- Understand how human activity affects a landscape.

Glaciated Upland Features

- River valleys become U-shaped or **troughs** with steep sides as a result of 'all-around erosion', which cuts off **interlocking spurs** to create **truncated spurs**. Many valleys contain **ribbon lakes**.
- Glaciated upland features can be seen in the Lake District, the north-west Highlands and Snowdonia.
- **Tributary** valleys were eroded less and remain high on valley sides. They may enter the trough as waterfalls.
- Ice collection hollows at altitude can become **corries** with steep, frost-shattered back walls, over-deepened floors and a convex exit or threshold.
- If corries develop close together, their back walls create an **arête**: a steep-sided, frost-shattered ridge. Three or more corries together make pointed summits or **pyramidal peaks**.
- **Roches moutonnées** are asymmetric rock features; they are abraded and smoothed **upstream** but rugged and plucked **downstream**.

> **Key Point**
>
> Both weathering and erosion processes contribute to the formation of features.

River flows between interlocking spurs

The ice carves away the land where the spurs interlocked

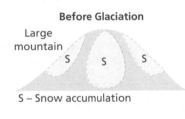

Before Glaciation

Large mountain

S — Snow accumulation

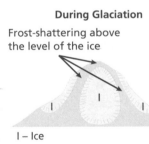

During Glaciation

Frost-shattering above the level of the ice

I — Ice

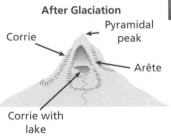

After Glaciation

Corrie
Pyramidal peak
Arête
Corrie with lake

Features of Glacial Deposition

- Terminal moraines are crescent-shaped ridges, often tens of metres high in the UK, made of mixed material and stretching across a glaciated valley. They mark the furthest limit of the ice.
- Ground moraine, or till, is a mixture of rock particles held in clay, smeared onto the surface and often very thick.
- Lateral moraines are ridges of debris along the side of the valley.
- **Drumlins** are rounded, elongated, asymmetrical mounds of moraine shaped by moving ice. They have a blunt 'stoss' upstream end and a tapered tail or 'lee' side on the downstream side. They usually occur in groups.

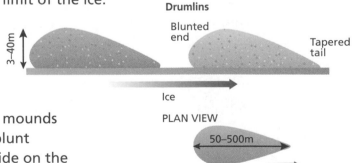

Drumlins

Blunted end
Tapered tail
3–40m
Ice

PLAN VIEW

50–500m
Ice

- Moraine is not solid rock so is easily remodelled by **meltwater**, weathering and erosion by rivers.

After Ice has Gone

- Lowland areas show evidence of both erosion and deposition.
- As ice retreated there were huge amounts of meltwater, which swept away loose material and carved out new river systems.
- Weathering continues to take place.
- Freeze-thaw action is mainly restricted to mountain regions in winter.
- Erosion and deposition by moving water creates new landforms and modifies others.
- Mass movement changes slopes.
- Human activities influence process and landscape.
- Some activities, especially road building, make slopes hazardous: steepening the slope angle; increasing the slope height; **saturating** the surface; increasing the weight on the surface.

A Lake District Valley Today

- The valley in the photo has been altered by time, natural processes and human activities.
- The lake bed is infilled by rivers, mass movement and slope processes.
- It has plucked rock faces, weathered and smoothed.
- The narrow, post-glacial river channel is winding across the valley floor.
- Vegetation has fixed scree slopes, reducing weathering and mass movement.
- **Drainage** ditches are cut to dry out arable land and aid flood protection.
- Evidence of farming: buildings on valley floor; field boundaries; improved soils; sheep grazing; drystone walls; a lack of trees on lower slopes is evidence of **overgrazing** by sheep.
- Evidence of tourism: scenery attracts visitors; opportunities for walking and climbing, winter and water sports. Results in slope erosion.
- Building roads can change slope angles, making them less stable.
- River banks strengthened and straightened to prevent flooding. However, this affects discharge patterns.
- Tree growth shows post-glacial climate improvement and soil creation.

Key Point

Today's landscapes result from rock type, structure, erosion, weathering and mass movement. The impact of major events like glaciation is modified over time.

Key Words

trough
interlocking spur
truncated spur
ribbon lake
tributary
corrie
arête
pyramidal peak
roche moutonnée
upstream
downstream
drumlin
meltwater
saturated
drainage
overgrazing

Quick Test

1. Where do arêtes form?
2. Name two features you might find along a U-shaped valley side.
3. Why does a roche moutonnée have a smooth upstream side?
4. Why does a drumlin have an elongated downstream side?

Coasts 1: Weathering and Erosion, and Associated Landforms

You must be able to:

- Describe weathering processes operating in coastal regions
- Describe mass movement at the coast
- Describe type and nature of waves
- Describe erosion processes
- Describe and explain landforms: cliffs, headlands, bays, wave cut platforms, caves, arches, stacks and stumps.

Weathering

- Weathering is the breaking down of rock without moving it away.

Physical

- **Frost action**: water trapped in cracks expands on freezing, weakens and breaks rock.
- **Salt crystal growth**: salt in seawater remains in cracks and dries, expands with further wetting, expands, weakens and breaks rocks.
- Plant roots and burrowing animals can put pressure on rocks and break them.

Chemical

- **Solution**: dissolving chemicals in rocks, especially limestones.
- Acids from decomposing **algae** aid decomposition.
- Seaweed growth breaks down rock and makes it easier to carry away.

Types of Mass Movement

- **Rock fall**: loose material detached from surfaces; associated with resistant rocks.
- **Sliding**: sudden movement of a large mass of surface material; moving at the same rate; collapses at cliff base into an unconsolidated mound.
- **Slumping**: sudden movement along a curved path; material has become too wet or heavy to stay attached to the cliff face.

Waves

- Waves are caused by wind. The **fetch** is the distance over which a given wind has blown. The longer the fetch and the stronger the wind, the larger the waves.
- Waves break against cliffs or in shallower water, releasing energy.
- **Swash** is running up the beach and **backwash** is going back.
- **Plunging** waves are powerful and bring material **seaward**.
- **Surging** waves push material up, making a ridge or higher beach.
- **Spilling** waves push material onshore.
- **Constructive**: swash has greater impact than backwash; material added and beaches get higher. 6–8 waves per minute.
- **Destructive**: backwash has greater impact than swash; material removed and beaches get lower. 10–14 waves per minute.

Key Point

Coastal landscapes result from the interrelationship of many factors: rock type and structure, the weather, vegetation, surface movements, and erosion by the sea.

Plunging Wave

Surging Wave

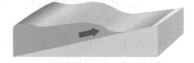

Spilling Wave

Constructive Wave

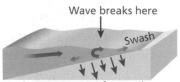
Wave breaks here
Swash
Large amount of water loss into the beach

Destructive Wave

Wave breaks here
Swash
Backwash
Small amount of water loss into the beach

Types of Erosion at the Coast

- **Hydraulic action**: wave quarrying; waves break, trapping air in joints, cracks, bedding planes; compression exerts huge force which breaks up the rock; loose debris falls to ground.
- **Abrasion**: waves use loose material to grind at the cliff foot and hurl it at the cliff face.
- **Attrition**: sediment crashing together when moved by waves; particles get smaller and rounder.

Attrition

Rock pieces detached from cliffs get smaller and rounder as they crash into each other moving up and down the beach.

Hydraulic Action

Breaking waves force air into cracks in a cliff, exerting great pressure.

Abrasion

Sand, pebbles and boulders carried by waves are hurled against cliffs, weakening and breaking the rock.

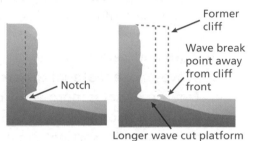

Landforms of Erosion

- **Cliffs**: form depends on rock type – more **resistant** rock, e.g. granite, gives higher, steeper cliffs; less resistant rock, e.g. clay, gives lower, gentler cliffs; surface weathering and mass movement; undercut by waves at the base which forms a **notch**; further erosion can cause collapse.
- **Wave cut platforms**: the result of cliff weathering and erosion; as cliffs retreat, more base rock is revealed; lowered mainly by abrasion; growth reduces the power of waves to erode cliffs.
- **Headlands** and **bays**: more resistant rock sticks out into the sea as headlands; less resistant or weakened rock is eroded back into bays; waves then attack headlands more; well-developed on discordant coasts.
- **Cave, arch, stack, stump**: destruction of headlands by waves:

Formation of Wave Cut Platform

Former cliff

Wave break point away from cliff front

Notch

Longer wave cut platform

> ## Key Point
>
> The work of the sea makes coastlines smoother; bays fill in and headlands wear back.

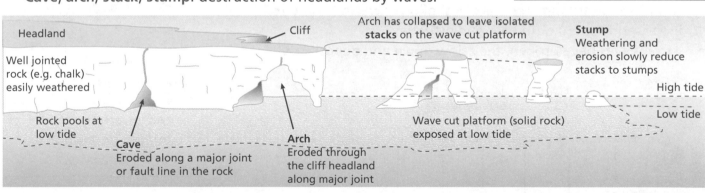

Headland

Well jointed rock (e.g. chalk) easily weathered

Rock pools at low tide

Cave
Eroded along a major joint or fault line in the rock

Cliff

Arch
Eroded through the cliff headland along major joint

Arch has collapsed to leave isolated **stacks** on the wave cut platform

Wave cut platform (solid rock) exposed at low tide

Stump
Weathering and erosion slowly reduce stacks to stumps

High tide

Low tide

Quick Test

1. What is the fetch?
2. What cliff shape is likely on hard rock?
3. Where would you see a notch?
4. What is a bay?

Key Words

frost action	slide	surging	abrasion	wave cut	cave
salt crystal	slump	spilling	attrition	platform	arch
growth	fetch	constructive	cliff	headland	stack
solution	swash	destructive	resistant	bay	stump
algae	backwash	hydraulic	notch		
rock fall	plunging	action			

Coasts 2: Transport and Deposition, and Landforms of Deposition

You must be able to:

- Describe how and why material is transported at the coast
- Explain why deposition takes place
- Describe landforms: beaches, sand dunes, spits and bars.

Coastal Transport and Deposition

- Material can be **transported** by **traction** and **suspension**:
  - Traction: dragging particles along the seabed and **beach**.
  - Suspension: within the water; not floating.
- Material is moved by **saltation**:
  - Particles bounce or jump across the surface as winds blow.
  - The impact of one particle can propel others.

Direction

- Forward and back:
  - Waves move material towards the cliff and back again.
  - Finer material is more easily carried by backwash.
  - **Attrition** also takes place.
  - This results in 'sorting'; smaller material towards the sea.
- **Alongshore**:
  - Waves carry material by **longshore drift**.
  - **Swash** brings particles up on to the shore at an oblique angle.
  - **Backwash** returns them in a direct line.
  - Many tonnes per year can be moved along a stretch of shore.

Deposition

- Waves deposit material when they have insufficient energy to carry it. Deposition takes place in shallower water (owing to increased friction) or in bays and estuaries.
- After waves have broken, they slow down and deposit.

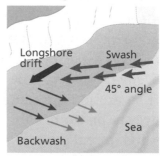
Longshore Drift

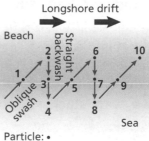

Particle: •

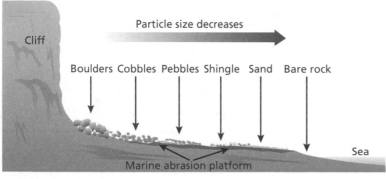

Landforms

- **Beaches** are usually composed of a similar type of material. Some contain mixed rocks left behind by glaciers.
- Debris lies on top of a wave cut platform.
- The biggest beaches form where waves approach parallel to the coast (i.e. 'swash aligned') and are mainly **constructive**.

> ### Key Point
>
> Beaches are the most common deposition feature at coasts.

This beach at Traighmhor, Lewis, is gradually getting higher and sand is covering the rocks

- Bay head beaches protect against more erosion and cliff retreat.
- A continuous supply of material is needed on coasts with a lot of longshore drift.
- The same beach will be a different shape throughout the year – higher in the summer owing to more constructive waves but lower in winter owing to plunging and destructive storm waves.
- **Sand dunes** are made up of sand blown to the back of a beach.
- Sand grains move by saltation and collect around an obstruction.
- A continuous, plentiful supply of dry sand is needed.
- Dunes form in lines, parallel to the shore. They are easily damaged but plants grow and help to stabilise them.
- Their colour changes from yellow to grey with age.
- Wind blows sand off the tops, which limits their height.

Key Point

There needs to be a continuous supply of material to be transported. This can be weathering debris, from offshore or alongshore.

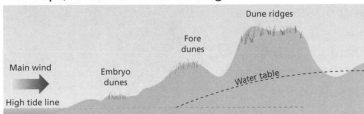

Waves rarely reach the back of a beach so the sand is dry and loose. Even between tides, it can dry enough to be bounced inland by an onshore wind.	**Embryo dunes** Sand grains move up the beach by saltation and get trapped by seaweed or debris. Plants, e.g. sea twitch or lyme grass, grow and bind the sand.

Fore dunes Tiny dunes join up to make yellow dunes. Made stable by marram grass.	**Dune ridges** Highest and biggest dunes. Look grey as plants die and decompose into the sand. Stop growing because as much sand is blown away as it arrives.

- **Spits** are long protrusions of sand and pebbles. They are usually close to an **estuary** and may have curved ends.
- Most material is brought by longshore drift, deposited and shaped by dominant waves. Some fine particles are brought from the estuary.
- The size of the spit is limited by other waves or by water flow from the estuary.
- The **proximal** end is attached to the land; the **distal** end is seaward.
- Calmer water inside a spit develops a **saltmarsh**.
- Spits can be **breached** in heavy storms then regrow.

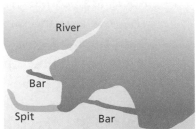

Sand bar at Padstow, Cornwall

1. Longshore drift brings sediment, which collects.
2. Finer sediment is collected in calmer water behind spits.
3. Storm waves hurl pebbles on to the spit, helping growth and stabilisation.
4. River flow stops growth.
5. Waves make the end of the spit curve inwards.

- **Bars** are made up of material deposited across a bay or parallel to the coast. They are often offshore of a gently sloping, sandy beach and usually only visible at low tide.
- Waves break and deposit material, then run up the beach.
- Bars can eventually form a complete barrier with a lagoon behind.

Quick Test

1. How does material change across a beach?
2. What process can move large amounts of material along a beach?
3. Where would you expect to find a saltmarsh?
4. What is traction?

Key Words

transportation	sand dune
traction	spit
suspension	estuary
beach	proximal
saltation	distal
alongshore	saltmarsh
longshore drift	breach
	bar

Coasts 3: Threats and Management

You must be able to:

- Understand that human use of the coast can conflict with natural processes
- Describe hard and soft engineering solutions to erosion
- Explain costs and benefits of management schemes
- Describe likely effects of climate change.

Human Use of the Coast

- Coasts are attractive places for **economic activity**, settlement, transport and **leisure**.
- They have **aesthetic** value.
- Construction on the coast adds weight and weakens rocks. Softer, unconsolidated materials are particularly susceptible.
- Human use of the coast can cause **conflict** with natural processes.

Hard Engineering

- **Sea walls** are built parallel to the coast and reflect, not absorb, wave energy:
  - There are different styles and shapes.
  - They are expensive to build and maintain (costing thousands of pounds per metre).
  - Most show serious damage after 30 years.
  - They are often ugly and can prevent access to the beach.
- **Rock armour (rip-rap):**
  - Rocks are pushed into and on to the cliff face.
  - The rock size and mass absorbs wave energy.
  - Gaps trap and slow down water, so material is less likely to be removed.
  - Rocks protect against damage from particles thrown by waves.
- **Gabions** consist of rocks in mesh cages.
- **Groynes** are built at right angles to the beach and are made of concrete, steel or wood:
  - They interfere with longshore drift.
  - If they are too short, material is taken offshore.
  - Places beyond groynes can be starved of sand.

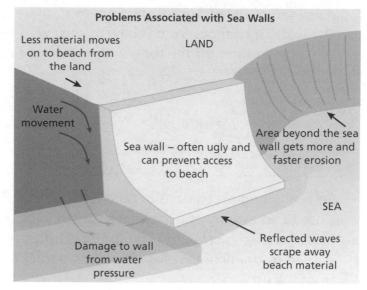

Problems Associated with Sea Walls

Less material moves on to beach from the land

LAND

Water movement

Sea wall – often ugly and can prevent access to beach

Area beyond the sea wall gets more and faster erosion

SEA

Damage to wall from water pressure

Reflected waves scrape away beach material

Sea Wall with Lid

Ramp Sea Wall

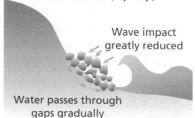

Rock Armour (Rip-Rap)

Wave impact greatly reduced

Water passes through gaps gradually

Groynes: Plan View

Area on downdrift side starved of sand

Groyne

Groyne

Groyne

Waves

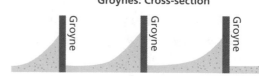

Groynes: Cross-section

Groyne

Groyne

Groyne

Soft Engineering

- Dune stabilising and **reprofiling**:
  - Involves stabilising dunes by fences built at seaward edge (sand accumulates around the fences).
  - **Marram grass** is planted above highest wave reach to stabilise the sand; **lyme grass** and **sea couch** are planted below.
  - Low cost but labour intensive.
  - Dunes are easily damaged in storms; reprofiling creates a new surface about 1 m above the highest wave run-up.
  - Material is transferred up or down the same beach so it automatically 'matches'.
  - Can involve repairing dune damage (blow-outs) or rebuilding longer lengths.
  - Speeds up dune recovery after storms.
  - Material can be from behind groynes or breakwaters.
  - Moderate costs but ongoing maintenance of thousands of pounds per 100 m.
- **Beach nourishment**:
  - Involves adding material to the existing beach so that wave energy is reduced.
  - Material should match the existing beach in size and type; it can come from offshore but will disturb wave patterns.
  - Moderate cost of £5000–200 000 per 100 m.
  - Material can be imported (but is more expensive).
  - Beach attractiveness can be improved.
- **Managed retreat**:
  - Involves the removal of coastal protection, usually in areas that were claimed from the sea.
  - It restores original sediment movements; beaches recover and grow; saltmarsh regrows.
  - Cheap (only initial removal of old structures needed).
  - It eventually creates natural defences for the **landward** area.
  - An example of managed retreat is Medmerry, West Sussex (2013).

Effects of Climate Change

- More severe and frequent storms.
- Coastal areas will be inundated by seawater.
- Existing defences could be overtopped.
- Worse impacts if river mouths are in the same area.

Example of Coastal Management: Sefton, Lancashire
- Largest dune system in Britain.
- Dunes soft and mobile.
- Over 5000 years old.
- Erosion and deposition.
- Natural and man-made defences.
- Management causes further problems.
- Varied history: Formby Point eroding now and in 18th century; advanced 300 m in 19th century.
- Pine trees planted to stabilise but affected light levels; altered ecosystem and sand washed away from roots. However, red squirrels (a protected species) established.
- Damage: beach activities, golf courses, walkers, sand extraction, farming behind dunes, settlements.
- 'Another Place' art installation attracts many visitors; revenue versus damage?
- Climate change predicted to raise sea level by 0.3 m.
- Stormier conditions will threaten much of this coast.

Key Point

Organisations attempt to modify behaviour to help protect coastlines.

Quick Test

1. Give another name for rip-rap.
2. What happens to wave energy at most sea walls?
3. Which are cheaper: soft or hard engineering strategies?
4. What happens if groynes are too short?

Key Words

economic activity	reprofiling
leisure	marram grass
aesthetic	lyme grass
conflict	sea couch
gabions	landward
groynes	

Rivers 1: Drainage Basins

You must be able to:

- Identify the components of a drainage basin
- Describe how water gets into a river channel
- Describe how water in channels is measured
- Understand the use of hydrographs.

Drainage Basins

- A **drainage basin** is an area of land whose rainwater feeds into one river channel.
- Drainage basins are separated by **watersheds**.
- Rivers flow from **source**, e.g. glacier; marshy uplands; lake; springs.
- Contributing streams are **tributaries**.
- Where two rivers meet is a **confluence**.
- Rivers enter the sea at the river **mouth**.
- Some drainage basins are tidal (estuaries).

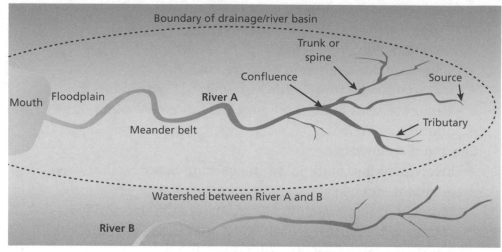

Parts of a River System

River Channels

- A **channel** is where a river flows at the base of a valley.
- The most efficient channel is semi-circular in cross-profile and twice as wide as it is deep.
- The **wetted perimeter** is the area touched by water.
- Water has to overcome friction to move.
- Changes to the channel affect the shape of the valley.
- The rate at which water reaches a river channel is affected by:

① **Geology** (the rock type):
  - **Permeable** rock (e.g. chalk) holds water.
  - **Impermeable** rock (e.g. granite) enables water to flow over the surface.

② **Relief** (the shape of the land):
  - Steeper slopes send water quickly into channels.

> ### Key Point
>
> Fast, intense rain is more difficult to absorb into the ground than slow, steady rain.

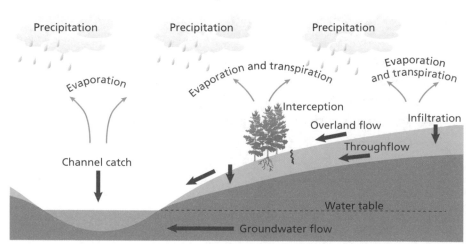

How Water gets into a Channel

3. Land use:
   - Settlements have impermeable surfaces so water flows quickly. Farmland absorbs water.
   - Drainage pipes remove water quickly.
   - Trees **intercept** rain and take up water from soil. Needle-leaved trees intercept more than broad-leaved in winter.
4. The amount of water in the land already:
   - Saturated ground becomes impermeable.
   - Baked-hard or frozen ground is impermeable.

- **Discharge** is the amount of water passing one place at one time and is measured in **cumecs** (cubic metres per second). It is calculated using channel cross-section and water velocity.

Routes into a Channel

	Where from	Quick or slow?
Channel catch	Rain going directly in	Quick
Overland flow	Valley side surfaces	Quick
Throughflow	Soil under valley sides	Slow
Groundwater flow	Rocks below the river	Very slow

KEY
...... Wetted perimeter
w width
d depth
v velocity

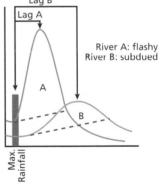

Cross-sectional area = w × d
Discharge = w × d × v

Hydrographs

- A **hydrograph** is a diagram showing the amount of discharge in one channel/river over a given length of time.
- Annual hydrographs run from October to September to show the wetter and drier seasons.
- Storm hydrographs show changing discharge after a rainfall event.
- Lag time is the time between highest rainfall and highest discharge and is the hydrograph's most important feature.
- Steep rising and falling limbs indicate a **flashy** river.
- Gentle rising and falling limbs indicate **subdued** rivers.
- The hydrograph shape is influenced by drainage basin characteristics as well as rainfall events:
   - Lots of tributaries move water quickly into the main channel.
   - Smaller drainage basins react more quickly to storms.
   - The basin shape also affects the hydrograph.

Key Point

The faster water can get into a channel, the more likely a river is to flood.

Example of a Hydrograph

Contrasting Hydrographs

River A: flashy
River B: subdued

Drainage Basin Shape

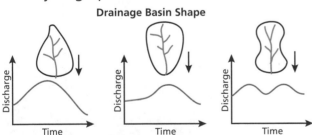

Key Words

drainage basin	wetted perimeter	throughflow
watershed	geology	groundwater flow
source	relief	discharge
tributary	intercept	cumecs
confluence	channel	hydrograph
mouth	catch	flashy
channel	overland flow	subdued

Quick Test

1. The place where two rivers meet is the _____.
2. What is lag time?
3. What is interception?
4. What measurements are needed to calculate discharge?

Rivers 2: Process and Landscape

You must be able to:

- Explain how rivers entrain, erode and transport
- Explain why rivers deposit
- Describe river landforms.

Entrainment and Erosion

- **Entrainment** is the wearing away and taking away of material as part of the process of erosion. This needs lots of energy.
- Short periods of fast flow erode more than months of lower flows.
- Vertical erosion deepens channels; lateral erosion widens channels; headward erosion lengthens channels.

Erosion Processes

Hydraulic Action, Abrasion, Solution and Attrition

- **Hydraulic action** is the force of moving water. Fast-flowing water disturbs loose material. River banks are weakened, undermined, then collapse. It is not effective on hard rock.
- **Abrasion** is when entrained material scrapes away the bed and banks. Small particles are created and these are easily carried.
- Most **downcutting** of channels results from abrasion.
- **Solution** is dissolved material in water. It is effective on limestones but most rocks are partially soluble.
- **Attrition** is when load particles crash into each other. Material gets smaller and rounder.

Transport and Deposition

- Entrained material is river **load**.
- More and bigger particles can be carried in faster water.
- **Suspended load** is carried within the water.
- **Bedload** is dragged along the channel floor by **traction** and bounced along by **saltation**.
- Soluble material is carried invisibly in solution.
- Deposition happens because there is insufficient energy to transport material.

Erosion Landforms

Interlocking Spurs and Waterfalls

- **Interlocking spurs** develop with a strong enough flow in small and highland streams.
- The streams mainly erode downwards by hydraulic action.
- Water flows around obstructions.
- Land sticking into the channels forms spurs. They 'interlock', restricting views up or down valleys.
- The characteristic V-shaped valleys are created.

Key Point

Coarser material is easier to entrain than fine particles but more difficult to carry.

 Hydraulic Action Attacking the Side of a Channel

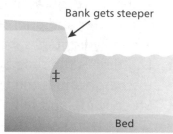

Bank gets steeper

‡

Bed

‡ Point of maximum speed and erosion

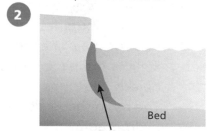

Bed

Material from the bank now available for river load

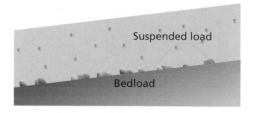

Suspended load

Bedload

Key Point

Attrition affects the load, not the channel.

- **Waterfalls** are the sudden steepening in the course of rivers. They occur on bands of different hardness of rock; where there is **faulting**; at **plateau** edges; with deposition in a channel; on the sides of glaciated valleys; and with changes in sea level.
- Moving water deepens the pool in front of the fall and erodes the rock behind by hydraulic action and abrasion, forming a **gorge**.

Waterfall Associated with Horizontal Bands of Rock of Different Hardness

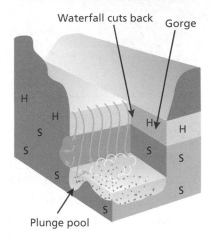

Waterfall cuts back   Gorge

Plunge pool

Waterfall Associated with a Fault

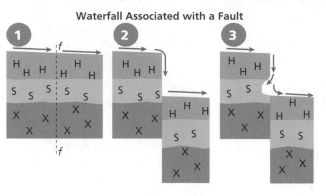

Waterfall Associated with Vertical Bands of Rock of Different Resistance

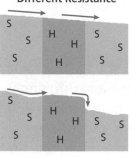

Erosion and Deposition Landforms

Meanders, Oxbow Lakes, Floodplains and Levees

- **Meanders** are extreme river bends with asymmetric cross-sections.
- Water takes a **sinuous** course, faster at one side than the other.
- Faster water hits the outer bend, and hydraulic action and abrasion undercut the bank, making steep river cliffs. Flow at the outer bend is **helicoidal**.
- At inner bends, water is slower so deposition takes place, forming a **slip-off slope**. At outer bends, a steep river cliff forms.
- **Oxbow lakes** are cut-off meanders.
- Erosion on meanders creates narrow necks. With very high water flow, the neck breaks through.
- Water initially flows through meanders and straightened sections. Further deposition leaves an oxbow lake.
- In a **floodplain, alluvium** is deposited at the sides of river channels.
- Coarser material is left from meander inner bends as they move.
- Floodwaters spread, depositing even very fine material.
- **Levees** are raised areas next to channels. Floodwaters suddenly slow and deposit material (coarser particles first).

Helicoidal Flow – Cross-section

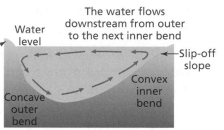

The water flows downstream from outer to the next inner bend

Water level

River cliff

Slip-off slope

Concave outer bend

Convex inner bend

Floodplains

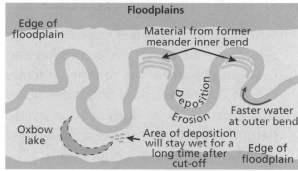

Edge of floodplain

Material from former meander inner bend

Deposition

Erosion

Oxbow lake

Faster water at outer bend

Area of deposition will stay wet for a long time after cut-off

Edge of floodplain

Levees

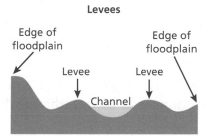

Edge of floodplain

Levee

Levee

Edge of floodplain

Channel

Quick Test

1. What is the river load?
2. Where are gorges formed?
3. Where does hydraulic action mainly attack?
4. Where are levees formed?

Key Words	
entrainment	plateau
hydraulic action	gorge
abrasion	meander
downcutting	sinuous
solution	helicoidal
load	slip-off slope
suspended load	oxbow lake
bedload	floodplain
faulting	alluvium
	levee

Rivers 3: Management and Flooding

You must be able to:

- Describe the use of rivers
- Describe the impact of humans on drainage basins
- Explain why rivers are managed
- Describe strategies of flood protection and management.

River Basin Use and Human Impact

- Human uses of rivers and valleys include drinking water, transport, **irrigation**, power, fishing, **recreation** and farming.
- Floodplains are used for settlement and industry.
- Water quality is threatened by human use, e.g. **nitrates** from farms; rainwater containing heavy metals flowing off roads.
- Groundwater input is reduced with increased use of impermeable urban surfaces.
- Buildings reduce bank stability.
- Construction erodes soil, which goes into channels.
- More storm drains, as a result of increased urbanisation, means that water gets into channels more quickly.
- Settlements cover river courses.

Management Strategies

- Straightening: increases speed of water in channels; increases erosion so the channel is kept deep enough for navigation.
- **Dredging** (clearing the channel): increases the channel size but velocity decreases, so is only a temporary measure.
- Levees (artificial versions of a natural feature): these raise the height of banks but create a problem with returning water to the channel after very high floods.
- Dams:
  - Advantages: reduce flow speeds to maintain river levels; control the amount of water being released; storage behind dams gives reliable water supplies throughout the year; HEP uses dam storage or regulated flow; boats and barges can use rivers all year.
  - Disadvantages: allow deposition behind; cause more erosion in front; change **ecosystems**; damage to surrounding rocks.

Flooding

- Natural rivers go over bank infrequently.
- Flooding is a useful natural process which distributes alluvium and creates fertile ground.
- Human activities, especially building, increase flood risk.
- Flooding causes expensive damage to property, short and long-term economic disruption and deaths in major incidents.

> ### Key Point
>
> Most human activities damage drainage basins and river systems.

> ### Key Point
>
> Management helps to ensure that a reliable water supply is maintained throughout the year and in droughts. It also helps to maintain irrigation of farmland, **hydroelectric power (HEP)**, **navigation** and flood protection.

Dam regulating a Pennine river

- Flooding is becoming a more frequent occurrence and is happening at different times; the main month for floods in the UK used to be August but recent serious events have been in the autumn or winter.
- Climate change predictions are for more frequent and bigger storms. Wetter winters mean an increase in severity, extent and frequency of flooding.

Flood Protection

Abatement

- Abatement involves tackling the problem at source, preventing water getting into channels. It includes attempts to lengthen lag times and reduce peak discharge.
- The channel or banks may be changed to take bigger discharges.

Hard and Soft Engineering

- Features of **hard engineering** include:
  - expensive costs but very rapid results
  - raising river banks to create bigger channels (but this can move the danger downstream)
  - building new relief channels to take extra water
  - diverting streams away from vulnerable areas
  - strengthening channels
  - dredging.
- Features of **soft engineering** include:
  - cheap costs but can take a long time to be effective
  - reducing slopes and terracing to reduce overland flow
  - tree planting to increase interception and soil water uptake
  - clearing boulders from river beds to maintain channel size and speed water through
  - using boulders to strengthen banks
  - discouraging building on floodplains to reduce the size of the threatened area and to minimise impermeable surfaces
  - earth **embankments** made from local materials; these can blend into the landscape but can get waterlogged or destabilised by plants and animals
  - 'wash' areas or flooding grounds to allow rivers to spread on to parkland and pitches
  - flood warnings, risk maps and preparation advice from the Environment Agency.

River banks reinforced with brick walls

Water level gauge and embankment on the River Ribble, Ribchester, Lancashire

> ### Quick Test
>
> 1. What is dredging?
> 2. Is reinforcing a river bank with bricks a hard or a soft engineering strategy?
> 3. Why do settlements increase impermeable surfaces?
> 4. What change to winter is predicted to happen with climate change?

Key Words
irrigation
recreation
nitrates
dredging
hydroelectric power (HEP)
navigation
ecosystem
hard engineering
soft engineering
embankment

Review Questions

Ecosystems and Balance

1. Give an example of a global large-scale ecosystem. [1]

2. Define the term 'producer'. [2]

3. Define the term 'consumer'. [2]

4. Define the term 'decomposer'. [2]

Total Marks _____ / 7

Ecosystems and Global Atmospheric Circulation

1. Which line encircles the Earth and divides it into the Northern Hemisphere and the Southern Hemisphere? [1]

2. Which line encircles the Earth at roughly 23° **north** of the Equator? [1]

3. Which line encircles the Earth at roughly 23° **south** of the Equator? [1]

4. Using examples, describe the global distribution of the tundra ecosystem. [4]

Total Marks _____ / 7

Ecosystems in the UK

1 What type of ecosystem can be found at Martin Mere in Lancashire and at Somerset Levels? [1]

2 Name one plant that is tolerant of water and salt. [1]

3 What type of ecosystem can be found in the Pennines of Yorkshire and in national parks in Devon? [1]

4 What two factors have made heathlands popular for farming and settlement? [2]

Total Marks _____ / 5

Tropical Rainforests

1 How have pitcher plants evolved to cope with the low nutrient soils in rainforests? [2]

2 Define 'ecotourism'. [2]

3 Apart from CITES and ecotourism, give two other ways in which rainforests can be sustainably managed. [2]

4 Using examples, describe how a named HEP scheme can have both benefits and drawbacks. [4]

Total Marks _____ / 10

Review Questions

Hot Deserts

1. Name a river that flows through the Mojave Desert in the USA. [1]

2. What two methods are used by farmers to improve the poor soil in the Mojave? [2]

3. Using an example, describe how plants have adapted to life in hot deserts. [3]

4. Using an example, describe how animals have adapted to life in hot deserts. [3]

Total Marks _____ / 9

Coral Reefs and Deciduous Woodlands

1. What word describes the climate in deciduous forests? [1]

2. How do deciduous trees increase water loss through transpiration? [1]

3. Give two ways in which the actions of humans can cause damage to coral reefs. [2]

4. Describe how coral reefs can be sustainably managed. [4]

Total Marks _____ / 8

Polar and Tundra Environments

1. What word describes the characteristic landscape type in tundra regions? [1]

2. Name two international agreements that control the human use of Antarctica. [2]

3. Using examples, describe how animals have adapted to life in polar and tundra climates. [4]

4. Name three resources found in and around Antarctica. [3]

Total Marks _____ / 10

Practice Questions

Glaciation 1: Rocks, Weathering and Mass Movement

1. Is frost-shattering a chemical or a physical weathering process? [1]

2. Explain why you would not usually expect to see surface water on a chalk hillside. [2]

3. Describe how metamorphic rocks are made. [3]

4. Why do you think south-east England is lower than north-west Scotland? [4]

Total Marks _____ / 10

Glaciation 2: Ice Processes

1. Where would you expect to see evidence of deposition by ice? [1]

2. Why is it important for material at the base of a glacier to be continually replaced? [2]

3. On the diagram below, which shows a glacier transporting material, what types of moraine are shown by A, B and C?

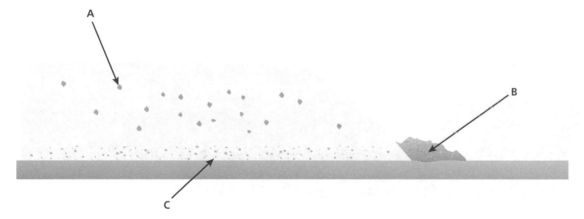

[3]

4. Explain the influence of the hardness of rock a glacier passes over. [3]

Total Marks _____ / 9

Practice Questions

Glaciation 3: Glacial, Physical and Human Landscapes and Post-Glacial Change

1. Where would you expect to find a terminal moraine? [1]

2. In what ways would the floor of a glacial trough look different today compared to just after the ice retreated? [2]

3. Explain the opportunities glaciated uplands present for tourism. [3]

4. How will soils on a U-shaped valley floor differ from those on the valley sides? [4]

Total Marks _____ / 10

Coasts 1: Weathering and Erosion, and Associated Landforms

1. a) Look at the diagram below, which shows mass movement on a cliff.

Cliff

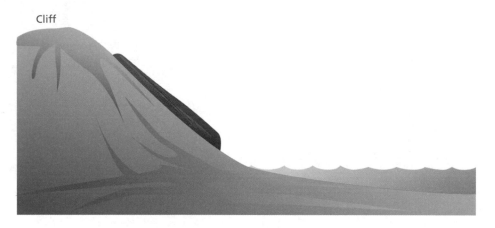

   What type of mass movement is shown? [1]

   b) What will happen to the surface material that is moving when it reaches the base of the cliff? [1]

2. On some coastlines, the rate at which the waves remove material is slower than the amount of material brought down cliffs by mass movement.

   In these cases, what will be the result for the coastline? [3]

Total Marks _____ / 5

Coasts 2: Transport and Deposition, and Landforms of Deposition

1 How is material moved along the shore by longshore drift? [4]

2 Describe and explain differences in winter and summer beach shapes. [4]

Total Marks _____ / 8

Coasts 3: Threats and Management

1 Look at this photograph taken in Newquay, Cornwall.

   a) Give one strength or positive aspect of this type of strategy. [1]

   b) Give one weakness or negative aspect of this as a form of coastal management. [1]

2 Using the diagrams below, describe some of the changes resulting from building at the coast.

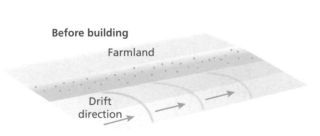

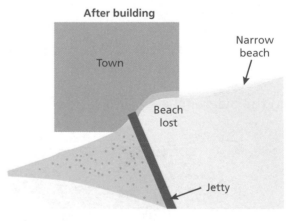

[4]

3 Why is a beach a good form of coastal protection? [3]

Total Marks _____ / 9

Rivers 1: Drainage Basins

1 Give two reasons why a river might be 'flashy'. [2]

2 Look at the storm hydrograph below.

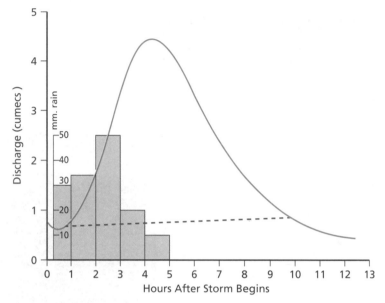

a) What was the lag time? [1]

b) What was the highest rainfall? [1]

c) What was the highest discharge? [1]

d) For how many hours was the river above normal flow? [1]

Total Marks _____ / 6

Rivers 2: Process and Landscape

1 On the diagram below, indicate where a waterfall is likely to form.

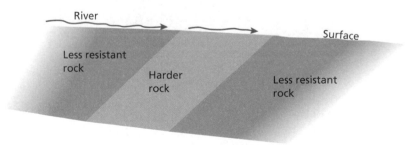

[1]

2 What will happen to the side of a channel affected by hydraulic action? [4]

3 Describe the formation of a waterfall resulting from faulting of rocks. [4]

Total Marks _____ / 9

Rivers 3: Management and Flooding

1 Look at the photograph below, taken in the Trough of Bowland in Lancashire.

a) What has been constructed on the river? [1]

b) Explain one reason why it might have been built. [2]

c) There are negative impacts of these structures – explain one. [2]

2 Describe two ways in which farming could damage the quality of river water. [4]

Total Marks _____ / 9

Urbanisation

You must be able to:

- Understand how rapidly the world's urban population is growing
- Describe how urban sprawl can be controlled
- Explain the reasons why people migrate to urban areas.

Urbanisation

- **Urbanisation** means an increase in the proportion of people living in urban (town and city) areas.
- In 1900, 10% of the world's population lived in urban areas. This had risen to 34% by 1960 and to 54% by 2015. The United Nations predicts that by 2030, 60% of the world's population will be urban.
- Levels of urbanisation are increasing across the world, with the fastest rates of growth occurring in lower income countries (LICs).
- As countries industrialise, levels of urbanisation increase as people move into cities for jobs.
- Most higher income countries (HICs) industrialised over 100 years ago and now have large urban populations. LICs are still in the early stages of industrialisation and so their urban populations are still growing.
- As urban areas grow outwards, they encroach into rural (countryside) areas. This is known as **urban sprawl**.

Urbanisation is increasing globally

Controlling Urban Sprawl in the UK

- Counter-urbanisation – the movement of large numbers of people from major urban areas into small towns or small villages – has been a growing trend since the early 1980s.
- This has created problems for these areas: newcomers often commute to work in the city, leaving empty 'dormitory villages' during the week; increased demand pushes up house prices so young local people cannot afford them; and villages and small towns grow and risk joining together, adding to urban sprawl.
- Greenbelts are bands of land surrounding major urban settlements on which no development can take place. They were introduced in 1947 to protect the countryside and control urban sprawl.
- New towns are brand-new planned settlements built on greenfield sites to accommodate the overspill population from existing large towns and cities where there is a shortage of housing.

Key Point

The majority of the world's population now live in urban areas.

World Millionaire and Mega-Cities

- A **millionaire city** is one with a population of over 1 million people. There are currently 280 millionaire cities in the world, and the number is growing rapidly – mostly in LICs.

Key Point

Most of the world's largest cities are in LICs.

- A city with over 10 million people is known as a **mega-city**. There are 35 of them in the world, mainly in LICs and newly emerging economies (NEEs).
- LIC cities are growing so fast due to rural to urban migration and high natural increase in population (high birth rates and falling death rates).

Rural to Urban Migration

- **Push factors** are those that drive people away from rural areas, e.g.:
  - unemployment
  - low wages
  - drought, famine and other natural disasters
  - farming is difficult and unprofitable
  - few job opportunities
  - lack of social amenities
  - isolation.
- **Pull factors** are those that attract people into cities, e.g.:
  - more job opportunities
  - higher wages
  - better schools and hospitals
  - better housing and services (like water, electricity and sewerage)
  - better social life
  - better transport and communications.

City life can attract people from rural areas

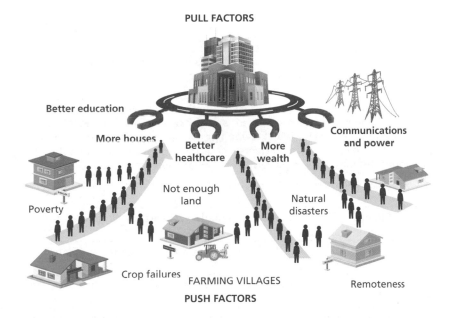

PULL FACTORS

Better education

More houses

Better healthcare

More wealth

Communications and power

Poverty

Not enough land

Natural disasters

Crop failures   FARMING VILLAGES   Remoteness

PUSH FACTORS

> **Key Point**
>
> LIC cities are growing rapidly, mainly due to rural to urban migration.

Quick Test

1. What is a millionaire city?
2. What percentage of the world's population was urban in 2015?
3. How many people are needed to qualify a city as a 'mega-city'?

> **Key Words**
>
> **urbanisation**
> **millionaire city**
> **mega-city**
> **push factors**
> **pull factors**

Urban Issues and Challenges 1

You must be able to:

- Explain how urban growth creates opportunities and challenges within cities in LICs and NEEs
- Explain different ways conditions in shanty towns can be improved for the residents.

Case Study: City in a Newly Emerging Economy (NEE) – Rio de Janeiro, Brazil

- Population: 12 million (2017).
- 23% of population live in over 600 **favelas**.
- Location: south-east coast.
- Reasons for growth: rural to urban migration and natural population increase.
- Former capital city of Brazil.
- Major tourist attractions: Copacabana Beach, statue of Christ the Redeemer, Sugar Loaf Mountain.
- Hosted football's World Cup in 2014 and the Olympic Games in 2016.

> ### Key Point
>
> Almost a quarter of Rio's population live in favelas.

Challenges in Rio

Social	**Migration:** Rapid growth in recent years because migration from rural areas to the city has put huge pressure on services and amenities. Migrants have been attracted to Rio due to job opportunities, higher wages, and better services like schools and hospitals.
	Housing: Rio has a number of shanty town settlements (favelas) that house the many rural migrants who come to the city. They are unplanned and spontaneous – often growing up on poor quality land. Residents have no legal land ownership. Houses are built using cheap materials like wood or corrugated iron and many people are crammed into a small area, leading to overcrowding, with no clean running water or sewage disposal. There are no schools or hospitals, few job opportunities and high levels of disease and illness.
	Healthcare: Has been poor but more clinics have now been established, especially to improve maternity care and support for the elderly.
	Education: Schools suffer from low enrolments, and funding has been made available for higher pay and better training for teachers, as well as grants for poor families.
	Water supplies: A clean water supply is not available for 12% of the population. New treatment plants have been built and new pipelines laid.
	Energy: A shortage of electricity has meant frequent blackouts. New power lines have been constructed, along with a nuclear power station and a new HEP station.
	Crime: Street crime has been high, with powerful drugs gangs controlling the favelas. The police have taken steps to reduce crime, including the 'Pacifying Units' that have reclaimed order in some of the favelas.
Economic	**Poverty:** There is a huge gap between rich and poor citizens in Rio. Some favelas have grown on hillsides right next to the affluent central business district.
	Employment: Unemployment remains high in the favelas (over 20%), where most people work in the informal economy. Poor transport systems make it difficult for favela dwellers to access other parts of the city.
Environmental	**Urban sprawl:** As the city continues to grow, it encroaches on surrounding rural areas.
	Pollution: Air pollution from heavy traffic and congestion is a major issue, as is pollution of the sea from sewage and industrial waste.
	Waste disposal: This is a particular problem in the favelas, many of which are inaccessible to collection vehicles.

Opportunities in Rio

Social	**Cultural and ethnic diversity:** Rio is home to a huge mix of different races, religions and cultures. The annual Rio Carnival is an international event.
	Education: Rio has a number of universities and is a centre for research and development in Brazil.
	Strength of community: There are many positive aspects to life in favelas, including the people who create their own economy, e.g. shops, restaurants and cottage industries (e.g. pottery); residents send money back home to families in rural villages; recycling commonly takes place, for example of waste and building materials.
	Transport: Economic development has led to improvements in roads and transport systems.
Economic	**Industry:** Industrial growth has boosted Rio's economy through steel making, port industries, oil refining, petrochemicals, manufacturing (including modern industries like computers and electronics) and a growing range of services such as banking and finance.
	Tourism: Rio is one of the most-visited cities in the Southern Hemisphere. Major attractions include Copacabana Beach, Ipanema Beach, the statue of Christ the Redeemer and Sugar Loaf Mountain.
Environmental	**Beaches:** The Atlantic beaches continue to attract tourists.
	Forests: The Tijuca National Park is one of the largest urban forests in the world.

Rocinha Shanty Town, Rio de Janeiro

- The rapid growth of Rio's population has led to a severe shortage of housing. The city has a number of shanty town settlements, such as Rocinha, that house the rural migrants who flock there.
- Rocinha is the largest shanty town in Rio, with a population of around 200 000 people. It is located in the south zone of the city on a steep hillside overlooking the city beaches.
- Not all the people in Rio are poor – many wealthy people live close to the central business district (CBD).

Improving the Quality of Life in Rocinha

- In the 1990s, the Favela Bairro Project (a slum to neighbourhood project) was set up to upgrade the favelas (as opposed to demolishing them) by providing pavements, electricity and sewage systems.
- The project promoted self-help building schemes, where local residents were provided with materials such as concrete blocks and cement to construct permanent dwellings, often with three or four floors and with water and sanitation.
- Residents were given legal rights of ownership or low rents on properties. Improved transport systems gave them better access to work in the city.
- Businesses like shops and restaurants were encouraged.
- Law and order improved through 'pacification' programmes.
- After the area was made safe for visitors, tourism flourished.

Key Point

Rapid urban growth of Rio has presented many social, economic and environmental challenges.

View over Rocinha favela

Key Point

Conditions in favelas can be improved with government support and self-help projects.

Key Words

favela

Quick Test

1. What percentage of Rio de Janeiro's population live in favelas?
2. Explain three ways in which living conditions in Rocinha favela have been improved.
3. Explain how self-help schemes have improved conditions in Rocinha favela.

Urban Issues and Challenges 2

You must be able to:

- Explain how urban changes in cities in the UK lead to a variety of social, economic and environmental opportunities and challenges
- Explain different ways inner city areas can be regenerated.

Case Study: UK City – London

- Population: 8.7 million (2015).
- Leading global city for industry, education and finance.
- A world cultural capital.
- The most-visited city in the world.
- Diverse range of peoples and cultures.
- Hosted the Olympic Games in 2012.
- During the 19th century, the port of London was the busiest in the world. By 1980, ships had become too large to sail up the River Thames and the docks in the heart of the city became derelict.

Challenges in London

Social	**Housing inequalities:** House prices in London are greater than anywhere else in the UK. Within the city, there is not enough good quality and affordable housing. Overcrowding has risen (particularly in the private rental sector) and is considerably higher than the rest of the country.
	Education: Overall attainment in London schools matches the rest of the UK, although there is considerable variation across boroughs in the city.
	Health: People with poorer English language skills have found it difficult to access healthcare facilities.
	Urban sprawl: As London has expanded, new buildings have been constructed on greenfield sites around the city.
	Migration: Job opportunities, particularly in finance and 'knowledge-based' industries, have attracted people to London from elsewhere in the UK and the rest of the world. It is estimated that up to one-third of all international migration into the UK is to London.
Economic	**Industry:** As the docks closed, many manufacturing industries were lost.
	Pay inequality: London has the most unequal pay distribution of any part of the UK, largely due to the high wages paid at the top end of the scale. 21% of people living in London are paid below the London living wage.
	Unemployment: The closure of factories led to high unemployment in parts of inner London.
Environmental	**Dereliction:** As manufacturing industries have declined (particularly in the eastern parts of the inner city and along the banks of the River Thames), much land has been left in a state of dereliction.
	Urban sprawl: This has increased pressure on land use in the rural–urban fringe and led to the growth of commuter settlements.
	Waste disposal: A large urban population produces a lot of household and commercial waste for disposal.
	Pollution: Atmospheric pollution from industry and vehicles.

Opportunities in London

Social	Diversity: Ethnic and cultural diversity allows people to experience different foods, music and religions.
	Entertainment and culture: London is an international centre for sporting events, theatres, cinemas, museums and art galleries.
Economic	Industry: The number of job opportunities in financial services and knowledge-based industries has increased. In addition, the redevelopment of London's docklands and the more recent redevelopment of the Olympic site in East London has increased the number and variety of jobs available in London.
	Transport: A new high-speed train link (HS2) from London to Birmingham and eventually on to Leeds and Manchester is planned. Work continues on the Crossrail project from Paddington station to Reading in the west and Abbey Wood to the east of the city. Plans have been approved to build a new runway to expand Heathrow Airport.
	Tourism: London is the most-visited world city and attracts tourists with its historic buildings, monuments, sports events, cultural events and entertainment.
Environmental	Reuse of industrial land: Numerous brownfield sites have become available for development.
	Transport: Improvements to transport systems aim to reduce carbon dioxide emissions by 60% by 2025. This includes wider use of diesel-electric hybrid buses, and the trialling of hydrogen fuel cell buses and combined diesel and biofuel buses.
	Regeneration: London's docklands have been regenerated, as part of which the Olympic site has been further developed. Other sites like Battersea Power Station and the Gas Works at Greenwich have also been regenerated for flats. The O2 Arena at Greenwich Peninsula has opened as a multi-purpose indoor arena.

Regeneration of London's Docklands

- The London Dockland Development Corporation (LDDC) spent £10 billion improving this part of London between 1981 and 1988.
- Disused dock basins like the Royal Docks were redeveloped with a mixture of luxury housing, hotels, shops and restaurants.
- New office development took place at Canary Wharf.
- The Docklands Light Railway (DLR) connected the area to the London Underground system.
- London City Airport – a STOL (short take-off and landing) airport.
- Affordable housing for local people.
- New cultural venues like the Docklands Arena and O2 Arena.
- New 'eco' housing, e.g. at Greenwich Millennium Village.
- The London Olympic site left a legacy of housing, sports facilities, improved transport links and the Queen Elizabeth Park.
- The Olympic Stadium is now the home ground for West Ham United.

 Key Point

London's docklands fell into decline but have been regenerated with different land uses.

The regenerated Victoria Docks in London's East End area, with the financial district in the background

Quick Test

1. List four modern uses for London's derelict dock areas.
2. Name two planned developments aimed at improving transport in London.
3. What new use was found for the Olympic Stadium after the 2012 Games?

 Key Words

brownfield site
regeneration

Urban Issues and Challenges 3

You must be able to:

- Explain why the UK population has grown over the years
- Describe a model to explain the different land uses found in cities
- Understand why urban areas need to be developed for the future in a sustainable way
- Explain how sustainable living involves a number of changes to current urban lifestyles.

Urban Change in the UK

- More and more people moved into UK cities from the beginning of the 1800s in search of jobs provided by the factories of the Industrial Revolution.
- At the same time, they were driven away from the countryside as mechanisation took over their jobs on farms. There was also surplus labour and an agricultural recession following the Napoleonic Wars.

Urban Models

- Geographers draw models to help them understand patterns like different land uses in cities.
- One of the most famous models was devised by Robert Park and Ernest Burgess, and is called the Concentric Zone theory.
- Another model is called the sector theory, and was devised by Homer Hoyt. It combines the rings of Burgess's model with wedges of land use that radiate from the city centre.
- These two models were combined by Peter Mann to form another version that applies particularly to British cities.
- **Central business district (CBD):** found in the centre of the city. Features include shops, banks and offices, high land values, tall buildings, high density of buildings and few residents.
- **Zone of transition:** found around the CBD, and containing a mix of residential use and older industrial buildings. This is an area in flux where land use is changing, with many derelict areas (some having been redeveloped or regenerated). This area is often referred to as the 'inner city'.
- **Industry / Factory zone:** area of older industrial buildings, many of which have closed down and fallen into disrepair.
- **Residential (suburbs):** these are housing areas increasing in modernity towards the edge of the city. Outer suburbs consist of large, detached houses with high prices.
- **Rural–urban fringe:** found on the very edge of the city where there are both urban and rural land uses, such as factories and out-of-town shopping malls mixed with agriculture.

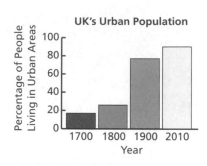

UK's Urban Population

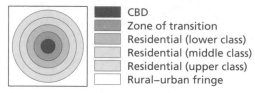

Concentric Zone Model (Park and Burgess)

CBD
Zone of transition
Residential (lower class)
Residential (middle class)
Residential (upper class)
Rural–urban fringe

Hoyt's Sector Model

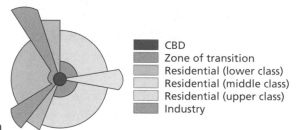

CBD
Zone of transition
Residential (lower class)
Residential (middle class)
Residential (upper class)
Industry

Mann's British Model

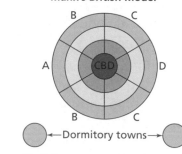

←Dormitory towns→

Age of building
▉ Transition zone
☐ Pre-1918
▉ Post-1918

Housing
A Upper⎫ middle class
B Lower⎭
C Working class
D Poorest houses/ industry

Key Point

Rapid change in many western cities, e.g. London, can make these models difficult to apply now.

Sustainable Urban Living

- In the UK, about 85% of people live in urban areas. Cities use huge volumes of resources and produce vast amounts of waste.
- However, urban areas can be developed in a **sustainable** way.
- Sustainable living means allowing people to meet their needs today without harming the prospect of people in the future to meet their needs.

Housing

- Ways to make housing more sustainable include:
  - low energy use, e.g. solar heating and efficient insulation
  - affordable prices and low rents
  - shared housing
  - passive housing: kept warm using heat from people, pets and natural lighting.

Transport

- Ways to make transport more sustainable include:
  - car-sharing schemes
  - vehicle restricted areas to reduce congestion and pollution
  - public transport (including buses, trains and trams) organised into an efficient integrated rapid transit system
  - alternative cleaner fuel sources, e.g. electric and hydrogen
  - cycle paths and walkways.

Energy Use and Waste Disposal

- The aim is to be '**carbon neutral**' – for a town to produce as much energy as it consumes. Some schemes that help include:
  - efficient recycling
  - reduction of household and industrial waste
  - composting of green waste
  - use of 'greywater' and rainwater collected for domestic use.

Services and Employment

- Daily needs like shops and schools should be within walking distance for the population.
- There should be a range of job opportunities for residents.

Environment

- 'Urban greening': a target of 40% green space, parkland and trees.
- Use of **brownfield** sites (once used for industry) for development.
- A greenbelt of natural space surrounding any settlement.

Quick Test

1. What percentage of the UK population lives in urban areas?
2. Describe the structure of Hoyt's Sector Model.
3. Identify three ways in which modern transport systems can be made more sustainable.
4. What does it mean to describe a sustainable city as self-supporting and self-contained?

 Revise

Key Point

Many current trends in urban living are not sustainable. Sustainable urban living includes methods of adaptation to, and mitigation of, climate change.

Key Point

Changes to housing and transport can make massive energy savings and help to manage climate change.

Key Point

Sustainable modern cities are self-contained, self-supporting and help to manage climate change.

Key Words

sustainable
carbon neutral

UK Population and Economic Change 1

You must be able to:

- Understand that the UK population is unevenly distributed
- Describe and explain the changes to the UK population structure in recent years.

Population Distribution in the UK

- Within the UK, there are areas with a relatively high population density like south-east England, and areas of low population density, like north-east Scotland.
- The major high density areas are the **conurbations** of London, Merseyside, Greater Manchester, West Midlands, West Yorkshire, Glasgow, Tyneside, Nottingham, Sheffield and Bristol. The borough of Islington (London) has a density of 15 179/km^2, whereas the Highland region (Scotland) has 60/km^2 (ONS, 2015).

How is the UK's Population Changing?

- The UK's population has reached stage 4 of the **Demographic Transition Model** (DTM, below right). See page 87 for more on the DTM.
- UK population rose rapidly between 1760 and 1880 as death rates fell while birth rates remained high.
- Population growth then slowed up as death rates continued to fall while birth rates fell rapidly.

Reasons for Declining Birth Rates

- Advances in contraception.
- More women in work and education and following careers.
- Improved healthcare reducing infant mortality.
- Financial restrictions on creating large families.

Reasons for Declining Death Rates

- Improved healthcare.
- New medical treatments and drugs.
- Fewer people in manual jobs.

UK Population Density

per km^2
- <1
- 1–25
- 26–100
- 101–250
- 251–1000
- >1000

Edinburgh
Glasgow
Belfast
Liverpool
Dublin
Manchester
Cardiff
Norwich
London

> **Key Point**
>
> The population of the UK is unevenly distributed.

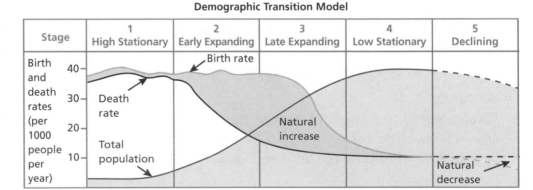

Demographic Transition Model

Stage	1 High Stationary	2 Early Expanding	3 Late Expanding	4 Low Stationary	5 Declining

Birth and death rates (per 1000 people per year)
Birth rate
Death rate
Total population
Natural increase
Natural decrease

> **Key Point**
>
> The population growth of the UK has slowed because of changes to birth and death rates.

Population Pyramids

- **Population pyramids** show a country's population structure.
- The UK pyramid has changed over time, showing:
  - an increase in the number of children aged under five years old
  - an increase in the number of young adults aged 20–25
  - a high percentage in the 60–70 age group (the post-war 'baby boomers')
  - an increase in very old people aged 80 or over.

Effects of an Ageing Population

- Additional pressure on health and welfare services.
- Increased demand for retirement homes and nursing homes.
- Economic costs, e.g. paying more state pensions.
- Tax rises for the working population to pay for additional services.

Ethnic Diversity

- There has been an increase in Asian and Black ethnic groups in the UK.
- The increase in the non-British white ethnic group is mainly from an increase in economic migration from Eastern Europe.

Changes and Challenges in Rural Areas

- Counter-urbanisation has led to people leaving cities to commute to work from smaller rural towns and villages.
- These dormitory villages have grown in size and number at the expense of the rural countryside.
- The increased commuter population has pushed up house prices beyond the reach of local people.
- The rural–urban fringe on the edges of towns has become a popular target for development.
- Second-home ownership has increased in many rural areas.
- Unemployment is high in many rural areas.
- Rural transport links have been threatened by government cuts.
- Agribusiness and non-agricultural diversification have changed the face of the farming industry, e.g. farm shops, tea rooms, antique barns, riding centres, go-karting, etc.
- The Government's Rural Development Programme provides financial support to farmers.

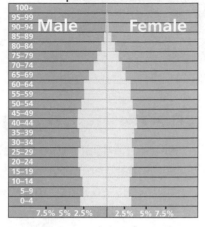

United Kingdom 2015:
Population 64 million

Key Point

The UK has an ageing population.

Key Point

Rural areas of the UK are being affected by population changes as much as urban areas.

Quick Test

1. Name four conurbations in the UK.
2. Outline two reasons for the UK's declining birth rates.
3. Outline two reasons for the UK's declining death rates.
4. List three challenges faced by UK rural areas as a response to a changing population.

Key Words

conurbation
Demographic Transition Model
population pyramid

UK Population and Economic Change 2

You must be able to:

- Explain the reasons for changes that have occurred to employment in the UK
- Describe the main features of a case study of a UK industry
- Understand the changes experienced by the retail industry in recent times.

Economic Change in the UK

- Since the 1980s there has been a decrease in primary and secondary employment, known as **de-industrialisation**, owing to increased mechanisation and automation, and competition from overseas.
- Since 2001 there has been an increase in tertiary employment, e.g. tourism, financial services and healthcare.
- Since 2001 there has also been an increase in quaternary employment – knowledge-based industries like software companies, research and development, and biotechnology.
- Quaternary employers are often based in science and business parks on the edges of major cities like Bristol and Cambridge.
- There has been an increase in flexible part-time working and home teleworking.

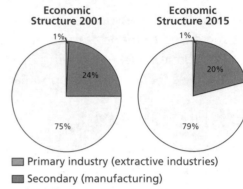

Economic Structure 2001
1%
24%
75%

Economic Structure 2015
1%
20%
79%

■ Primary industry (extractive industries)
■ Secondary (manufacturing)
□ Tertiary (services) and quaternary (IT and Research and Development)

New Developments in UK Infrastructure

- Road building and improvement programmes, including 'smart' motorways.
- Upgrades to the rail network, including the proposed HS2 link between London and Manchester/Leeds.
- UK airports have expanded and a third runway at Heathrow Airport is proposed.

> ### Key Point
>
> There has been a shift away from traditional forms of employment in recent years.

UK Case Study: High-tech Industries in the M4 Corridor

- New industrial regions in the UK have tended to develop along main communication routes.
- A good example of this is the M4 corridor, which also takes in the area around motorways such as the M11, M23 and M3.
- The area stretches from Heathrow Airport in the east to Bath and Bristol in the west, and is home to major international companies such as Hewlett-Packard and Sony.
- The industries found in this area can be described as **footloose industries**, as they are not specifically tied to one location.
- The M4 corridor has many locational advantages:
  - The motorway system allows easy access to all parts of the UK.

- Easy access to the Channel Tunnel and ports to allow exports abroad.
- Close proximity of Heathrow, Gatwick, Stanstead and Luton airports for international exports.
- A skilled workforce found in the university towns of Oxford, Cardiff, Reading and Bristol.
- Nearby major universities for expertise and research.
- Good access to London (seat of government and major financial centre).
- Other industries nearby mean that ideas, knowledge and services can be shared.
- Land on the edge of large towns and cities is cheaper.
- Attractive natural environment (e.g. Cotswolds and Mendip Hills) provides a pleasant place to live for the workforce.

Changes in Retail Provision in the UK

- Increased car ownership means people can travel further to out-of-town shopping centres.
- Growth of suburban shopping malls and business parks.
- Fewer small convenience stores survive in rural areas.
- Small specialist shops have given way to large chain stores and supermarkets where customers can access a large range of goods at cheaper prices. This has led to the growth of **'clone towns'**.
- Shoppers have less time to shop so make fewer visits to larger stores where they purchase more.
- Increase in 'e-tailing'; using the Internet.
- 'Ethical' shopping has become more popular (e.g. food provenance and food miles).

A modern shopping centre

Key Point

The retail industry has experienced many recent changes.

Issues Connected to Leisure Provision in the UK

- Traffic congestion at historical and cultural urban sites (e.g. Bath and Oxford).
- Land use conflicts within National Parks, e.g. between farmers and conservationists.
- Overcrowding at rural **'honeypots'** (e.g. Lake Windermere).
- Erosion to rural footpaths and trails.
- Litter in the countryside.
- Seasonal employment of some leisure industries.

Footpath erosion

Quick Test

1. What percentage of UK workers had jobs in the tertiary or quaternary sectors in 2015?
2. Identify three knowledge-based industries that have seen recent growth.
3. Describe three recent changes experienced in the retail industry.
4. Outline three issues connected to leisure provision in the UK.

Key Words

de-industrialisation
footloose industry
clone town
honeypot

Global Development 1

You must be able to:

- Understand that there is more to development than just wealth
- Identify reasons why some countries are developing slowly
- Explain the consequences of uneven world development.

Classifying World Development

- Development refers to the progress of a country in terms of economic growth, human welfare and the use of technology.
- The development of different countries can be classified as:
  - **LIC** (lower income country) – classified by the World Bank as less than $1045 GNI (gross national income) per capita.
  - **HIC** (higher income country) – classified by the World Bank as more than $12 746 GNI per capita.
  - **NEE** (newly emerging economy) – a country that has begun to experience high rates of economic development, usually with rapid industrialisation. They differ from LICs as they no longer primarily rely on agriculture. The leading NEEs are known as BRICS (Brazil, Russia, India, China and South Africa).

Measuring Development

- There is more to development than just wealth.
- Geographers use a variety of data (or indicators) to decide how developed a country is. These indicators can cover social, economic or environmental factors such as those shown to the right.
- One single factor does not always tell a full and accurate story about a country. It has been popular in recent times for geographers to use the **Human Development Index** (HDI), which combines life expectancy, literacy and income to measure development.

Human Development Index

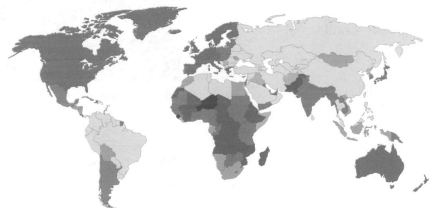

	High human development	0.9–1.00
		0.8–0.89
	Medium human development	0.7–0.79
		0.6–0.69
		0.5–0.59
	Low human development	0.4–0.49
		0.3–0.39
		0.2–0.29
	Not applicable	

Limitations of Economic and Social Measures

- The use of data can give a false picture of development as it gives an average for a whole country.

> **Key Point**
>
> Countries develop at different rates and there is a wide range of reasons for this.

> **Indicators used to determine a country's development**
>
> - GNI (gross national income) and GDP (gross domestic product)
> - Life expectancy (expected number of years of life)
> - Literacy (percentage of population that can read and write)
> - Birth rate (number of babies born per 1000 people per year)
> - Death rate (number of deaths per 1000 people per year)
> - People per doctor (population total divided by number of available doctors)
> - Calorie intake (total food calorie value per day)
> - Car ownership (number of cars owned per 1000 people).
> - Access to safe water (measured as a percentage of the total population)
> - Infant mortality (number of babies that die before one year of age, measured per 1000 live births each year)

> **Key Point**
>
> There are many ways to measure development, apart from wealth.

- The use of a single measure of development can be misleading, e.g. low birth rates generally suggest good social development, but birth rates can be reduced by government policy.
- Data can be out-of-date, unreliable or not available.
- The measures used may not take into account important aspects of development in a country, e.g. subsistence agriculture.
- Corrupt governments may distort the available data.

Population Characteristics of HICs
• Growth rate stable • DTM stage 4 • Low birth rates, low death rates • Ageing population • Some countries have declining populations (DTM stage 5)

Population Characteristics of LICs
• Population rising rapidly • DTM stage 2 or 3 • High birth rates, falling death rates • Many young dependants

The Demographic Transition Model

- The **Demographic Transition Model** (DTM) shows population changes over time. As a country becomes more developed, its population characteristics change.
- The DTM is often divided into four or five stages. Most HICs are in stage 4 or stage 5, while most LICs are in stage 2 or stage 3.
- The difference between birth rates and death rates represents the natural population increase. If birth rates are higher than death rates, the total population will increase.
- Global population is rising rapidly because birth rates are still very high in LICs, while death rates have fallen across the world.

Demographic Transition Model

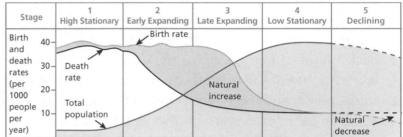

Causes of Global Inequalities

Physical causes	• Extreme climates making it difficult to produce food • Unproductive land for farming, e.g. too high, too steep, thin soils • Limited water supplies • Frequent natural hazards, e.g. tropical storms, earthquakes, tsunamis • Limited natural resources, e.g. iron ore, precious metals, energy sources
Economic causes	• Rapid population growth putting pressure on resources • Trade imbalance and reliance on primary resources for exports • International debt
Historic causes	• Corrupt governments • War and conflicts

The Sustainable Development Goals
The United Nations has identified 17 **Sustainable Development Goals** that came into place from January, 2016. Uneven world development means that some countries are not meeting these and are seeing worsening, not improving, living conditions: ❶ No poverty ❷ No hunger ❸ Good health ❹ Quality education ❺ Gender equality ❻ Clean water and sanitation ❼ Renewable energy ❽ Good jobs and economic growth ❾ Innovation and infrastructure ❿ Reduced inequalities ⓫ Sustainable cities and communities ⓬ Responsible consumption ⓭ Climate action ⓮ Life below water ⓯ Life on land ⓰ Peace and justice ⓱ Partnerships for the goals

Consequences of Uneven Development

- Disparities in wealth (the most developed countries have the greatest wealth and a good balance of trade; there can be wide differences in wealth within single countries, e.g. China).
- Disparities in health (LICs have a shortage of safe, clean water and are unable to invest in good quality healthcare, with infectious diseases such as malaria, tuberculosis and diarrhoea the main cause of deaths – particularly among children).
- International migration (both forced and voluntary, as people seek to improve their quality of life).

Key Words

Key Words
LIC HIC NEE **Human Development Index** **Sustainable Development Goals**

Quick Test

1. Who are the BRICS?
2. Identify three development indicators (other than wealth).
3. Name four of the global Sustainable Development Goals.

Global Development 2

You must be able to:

*   Understand how globalisation has connected different countries in the world
*   Explain the reasons for rapid development in one NEE
*   Describe a number of different categories of international aid.

Globalisation and Development

*   **Globalisation** refers to the process of countries of the world becoming interconnected and dependent on each other due to trade and improvements in communications, such as air travel and Internet links.
*   The general pattern of globalised world trade is one where LICs export raw materials (e.g. rubber, rice and oil) to HICs, and buy back manufactured goods. As the value of the raw materials is less than that of the manufactured goods, many LICs suffer a trade deficit, and this helps to widen the world development gap.
*   Some large industries are known as **transnational corporations (TNCs)**, as they produce goods and sell goods in a number of countries across the globe. Examples include Sony, Nike and Coca-Cola.
*   TNCs often locate their factories in LICs because labour is cheaper, working hours are longer, and there are fewer health and safety laws. However, the headquarters and research centres of TNCs are located in HICs.
*   TNCs can bring many positive benefits to the host country, such as more job opportunities, higher wages and provision of education and training.
*   TNCs also spend a lot of money improving infrastructure, e.g. roads and railway networks.
*   Local companies can also benefit by supplying to the TNC.
*   TNCs can also bring negative effects:
    *   The jobs created are not always secure.
    *   Factory employees often have to work long hours in poor conditions.
    *   Some notorious 'sweatshops' are filled with child workers.
    *   Factories have few controls on waste and pollution.
    *   Profits made by TNCs do not stay in the host country.

Case Study of Rapid Development: Vietnam

*   Country in South-East Asia (borders with China, Laos and Cambodia), with a population of 91 million (2014).
*   Newly-emerging economy; 7% annual growth (January, 2016).

Key Point

Countries of the world have become interdependent due to globalisation.

- Despite rapid growth, the majority of people are still working in primary sector jobs (56%).
- The secondary sector (43% employment) generates the most income, mainly because it has become an important offshore location for TNCs due to cheap labour, a growing home market and fewer industrial laws and restrictions.
- Nike has 34 plants in Vietnam (75% of its workforce is based in South-East Asia); 75 million pairs of trainers are made each year and the majority of workers are women under the age of 25.
- Advantages of TNCs: development of a widening industrial base; good wages; new skills; a positive contribution to GDP; extra tax money to spend on infrastructure; this all leads to the attraction of other TNCs.

Revise

Key Point

Countries like Vietnam are emerging as newly industrialised nations.

International Aid

- **Long-term aid** tries to solve a problem so that it never happens again, for example the construction of hospitals and a healthcare system to reduce mortality rates, or dam construction for HEP, regular water supplies and downstream flood control.
- **Short-term aid** tries to solve an immediate problem like the shortages of homes following an earthquake. It is sometimes called 'emergency aid' or 'crisis aid'.
- Government aid is money pledged by a country and raised from its tax system.
- Charity aid is funded by voluntary contributions from the public. Charity groups (or NGOs – non-governmental organisations) include Shelter, Oxfam and Comic Relief.
- Bi-lateral aid is an aid programme agreed between two individual countries.
- Multi-lateral aid consists of an aid package put together by a group of countries like the European Union or United Nations, perhaps to respond to a natural disaster.
- Tied aid (or 'conditional aid') means that the donor country demands something in return from the recipient country – often in some form of trade agreement.
- **Intermediate technology** (or 'appropriate technology') is a form of aid that improves existing simple technology, but does not necessarily use the most up-to-date technology to replace it. High-technology solutions may not be appropriate in certain situations due to the costs involved in purchasing, buying fuel and paying for maintenance.

Aid can fund pumps to provide clean water

Key Point

Aid can assist the development of LICs in a number of different ways.

Key Words

globalisation
transnational corporation (TNC)
long-term aid
short-term aid
intermediate technology

Quick Test

1. How has the pattern of globalised world trade helped to widen the development gap?
2. Identify three reasons why a TNC might want to locate in a country like Vietnam.
3. Explain the difference between long-term and short-term aid.

Review Questions

Glaciation 1: Rocks, Weathering and Mass Movement

1 How would you classify a slope angle of _8°_? ~~Weathering~~ Phy   [1]

2 What is the most common type of rock in Britain?   [1]

3 Give two differences between a flow and a slide.   [4]

4 Give two ways in which water is important in the weathering of rocks.   [4]

Total Marks _____ / 10

Glaciation 2: Ice Processes

1 Which has greater powers of erosion and transport: a river or a glacier?   [1]

2 What evidence might there be that abrasion by ice has taken place on a rock surface?   [2]

3 When would a glacier deposit material?   [2]

4 On the diagram below, insert an arrow to indicate the direction of ice flow and add the following labels in the correct places:

- Bedrock

- Plucking

- Debris

- Plucked rock

- Abraded surface

[6]

Total Marks _____ / 11

Glaciation 3: Glacial, Physical and Human Landscapes and Post-Glacial Change

1 Why are features of glacial deposition often missing from a present-day glacial valley? [2]

2 What role do you think abrasion plays in the formation of a corrie? [4]

3 What changes to the valley side slopes of a trough could have occurred since the end of the Ice Age? [3]

4 What economic opportunities are presented by an upland glaciated landscape? [4]

Total Marks _____ / 13

Coasts 1: Weathering and Erosion, and Associated Landforms

1 Describe the effect of hydraulic action on a cliff face. [2]

2 On some coastlines, the rate at which the waves remove material is faster than the amount of material brought down cliffs by mass movement.

In these cases, what will be the effect on the coastline? [3]

Total Marks _____ / 5

Coasts 2: Transport and Deposition, and Landforms of Deposition

1 Why are sand dunes easily damaged? [2]

2 Why are beaches temporary features? [3]

3 How is sand accumulated to make sand dunes? [4]

Total Marks _____ / 9

Review Questions

Coasts 3: Threats and Management

1 The photographs below were taken in locations close together on the Lancashire coast.

a) What damage is evident in the photograph on the left? [1]

b) What natural process of dune stabilisation can be seen in the photograph on the left? [1]

c) What has been done in the photograph on the right to help prevent further damage? [1]

d) Explain ways in which this method might be successful and why it might not. [4]

2 Describe and explain how sea walls work. [3]

Total Marks _____ / 10

Rivers 1: Drainage Basins

1 **a)** What is meant by the source of a river? [1]

b) Give two examples of sources. [2]

2 Describe two ways in which rainfall could be delayed or prevented from reaching a river channel. [4]

Total Marks _____ / 7

Rivers 2: Process and Landscape

1 Look at the photograph of a river in Scotland.

Location X

A

B

C

a) Name the features labelled A and B. [2]

b) What feature is likely to form at C? [1]

c) What part of the river is at location X? [1]

d) What process is operating at location X? [1]

2 **a)** Where does both erosion and deposition take place at the same time in a river? [1]

b) Why can erosion and deposition take place at the same time? [4]

3 Describe the formation of an oxbow lake. [4]

Total Marks _____ / 14

Review Questions

Rivers 3: Management and Flooding

1 Look at the photographs below taken on farmland near the River Hull in Yorkshire.

The soft engineering strategy of an earth embankment is visible.

a) Give one advantage of earth embankments. [1]

b) Give one potential problem with this strategy. [1]

c) Describe an environmental benefit that can occur with earth embankments. [2]

d) In urban areas, what are embankments usually made of? [1]

e) How do embankments help to prevent flooding? [2]

2 For a named place in the UK, describe **two** natural reasons why it is susceptible to flooding. [4]

Total Marks _____ / 11

Urbanisation

1 What percentage of the world's population lived, or is predicted to live, in urban areas in the following years?

Choose from the following possible answers:

10%   25%   42%   54%   60%

a) 1900   [1]

b) 2015   [1]

c) 2030   [1]

2 What does the term 'urban sprawl' mean, and why might it be considered a problem?   [3]

3 a) List some of the different push factors that lead to rural to urban migration.   [3]

b) List some of the different pull factors that lead to rural to urban migration.   [3]

4 Explain the differences in urban growth between the richer parts and poorer parts of the world.   [4]

Total Marks _____ / 16

Urban Issues and Challenges 1

1 Explain why so many rural people are attracted to cities like Rio de Janeiro in Brazil.   [4]

2 Rio de Janeiro has a large number of shanty towns.

What problems are faced by the people who live in them?   [4]

3 Tourism is a major industry in Rio de Janeiro. Describe some of the attractions the city has for visitors.   [4]

4 Describe in detail two social challenges of life in Rio de Janeiro.   [4]

Total Marks _____ / 16

Urbanisation Issues and Challenges 2

1 Why did the docklands in east London fall into decline in the 20th century? [2]

2 Identify four social improvements that formed part of the legacy of the London Olympics. [4]

3 Explain two ways in which public transport systems can be improved in London to help reduce carbon emissions. [2]

4 Describe two new projects that aim to improve transport networks in London. [4]

Total Marks _____ / 12

Urbanisation Issues and Challenges 3

1 What is the difference between a 'greenfield' and a 'brownfield' development site? [2]

2 Explain three ways that housing can be designed to be sustainable. [3]

3 Describe the characteristics of a central business district (CBD) in a typical UK city. [4]

4 According to the Concentric Zone Model, how does land use in a UK city change as you travel from the centre to the outskirts? [5]

Total Marks _____ / 14

UK Population and Economic Change 1

1 What is a population pyramid? [1]

2 What characteristics of birth rates and death rates are demonstrated in stage 2 of the Demographic Transition Model (DTM)? [2]

3 What problems are caused by Britain's ageing population? [4]

4 Describe how the structure of a country's population changes as it moves from stage 2 to stage 4 of the Demographic Transition Model (DTM). [3]

Total Marks _____ / 10

UK Population and Economic Change 2

1 Name two cities where science parks have developed. [2]

2 Why have out-of-town shopping centres (retail parks) grown in popularity in recent years? [3]

3 What is meant by the term 'ethical shopping'? [4]

4 Many footloose industries have grown up in the M4 corridor. Giving examples, explain why this location is so attractive to these industries. [6]

Total Marks _____ / 15

Global Development 1

1 Name two social indicators that could be used to measure development. [2]

2 Name two economic indicators that could be used to measure development. [2]

3 Name two environmental indicators that could be used to measure development. [2]

4 Explain how improvements in literacy can improve quality of life in LICs. [6]

Total Marks _____ / 12

Global Development 2

1 Why is the most up-to-date technology not always the best solution to development problems in LICs? [2]

2 What improvements in communication systems have led to the globalisation of businesses across the world? [3]

3 What is a 'TNC'? Name three examples. [4]

4 Explain how a TNC can affect local people and the economy of an LIC. [6]

Total Marks _____ / 15

Overview of Resources – UK

You must be able to:

- Explore how agricultural production in the UK is commercially organised and highly mechanised
- Explain how water is increasing in demand in the UK
- Explain how the sources of energy used to power the UK are changing.

Agriculture and Food in the UK

- Agriculture in the UK takes up around 70% of the land area, employing less than half a million people on over 200 000 farms, and contributes £10 billion to the economy – less than 1% of GDP.
- The UK produces less than 60% of the food it eats; the rest comes mainly from its European partners.
- Other products are imported from further afield, especially from countries like Kenya and Peru where the climate allows vegetables such as green beans, mangetout and peas to be grown all year.
- Agriculture occurs in most rural areas: the main concentrations are **arable** farming (crops) in the south and east of the country and **pastoral** farming (livestock) in the north and west.
- This is because of climatic factors: the south and east tend to be warmer and drier than the north and west, and also tend to have more fertile soils.
- Only around 25% of all the fruit and vegetables consumed in the UK are grown here.
- Most organic produce sold in UK supermarkets is grown within the UK but is only sold when 'in season', such as asparagus.
- The import of fruit and vegetables from countries like Spain, Kenya and Peru leads to increased **food miles** and **carbon footprints** but does help the economies of those countries.
- The increasing use of polytunnels can help to lengthen the growing season, especially in southern counties of England.

> **Key Point**
>
> UK agriculture is highly organised but still cannot produce enough to cover all of the country's food needs.

Water Demand and Usage in the UK

- With increasing affluence, water demand is growing as new houses are built with more than one bathroom and consumers demand labour-saving devices, such as dishwashers.
- Industry and agriculture are also large users of water, e.g. 15 000 litres of water are needed to produce 1 kg of beef.
- Maintaining the quality of drinking water and the management of water pollution are costly, requiring considerable investment in technology and with costs that are largely met by the consumer.
- The areas of highest demand in the UK – London and the South East – are the areas where there is least available water.
- Areas with less demand for water – the north and the west of the UK – are also the wettest and have surplus water.

- To maintain supplies, **water transfer schemes** deliver water from wetter to drier areas with a large demand. For example, water is moved south from Kielder Water in Northumberland to the cities of Yorkshire using a system of pipelines, aqueducts and rivers.

Energy and Demand in the UK

- The reliance on fossil fuels has a significant impact on the UK's energy security as increasing amounts of energy are being imported because domestic supplies of coal, gas and oil are running out.
- Around half of the UK's electricity is made from gas, the rest from coal, nuclear and renewables.
- The use of renewables is growing in significance as the number of wind and solar power installations increases, but they are less efficient in providing energy than fossil fuels.
- New methods of exploiting energy sources include shale gas that is obtained through the process of **fracking**, a hydraulic fracturing process that has many environmental issues, such as pollution of groundwater and the potential for earth tremors.

Key Point

Water demand is increasing in the UK, especially in those areas where water supply is low.

Key Point

The UK's reliance on fossil fuels is diminishing but is unlikely to disappear completely.

Fracking

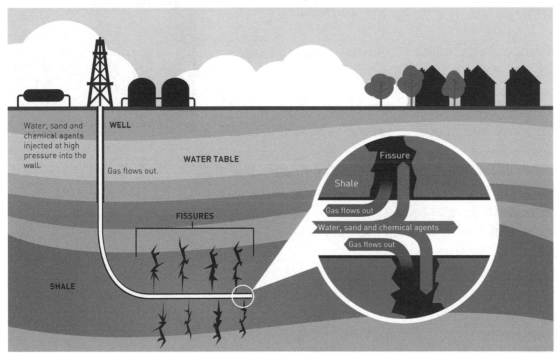

Water, sand and chemical agents injected at high pressure into the well.

WELL

WATER TABLE

Gas flows out.

FISSURES

SHALE

Fissure

Shale

Gas flows out

Water, sand and chemical agents

Gas flows out

Quick Test

1. Why does the UK need to import food from abroad?
2. Why is water demand increasing in the UK?
3. Why do London and the South East have problems in supplying enough water?
4. What challenges does the UK face to provide enough energy in the future?

Key Words

arable
pastoral
food miles
carbon footprint
water transfer scheme
fracking

Overview of Global Inequalities

You must be able to:

- Describe what natural resources are and how they are classified
- Explore patterns of resource consumption
- Explain how governments encourage efficient use of energy.

Global Supply and Consumption of Natural Resources

- **Natural resources** are material sources of wealth in the form of minerals, energy sources (**abiotic**), agricultural products and other food sources (**biotic**) that have economic value and can be traded.
- Europe, the USA and Japan generally consume the most resources per person.
- The regions that consume the least resources per person are Africa, South America and parts of Asia.
- Some of the richest regions have few natural resources (e.g. Japan), while some of the poorest regions are resource rich (e.g. Sudan).
- The richest 20% of the human population consume over 86% of all global resources, while the poorest 20% consume only 1.3% of all global resources.
- **Fossil fuels** such as oil, gas and coal are found in many countries around the world and consumed in every country. They are known as non-renewable fuels.
- There are big differences in fossil fuel consumption between different developed countries, and generally developed countries consume more than developing countries.
- As a country develops, the demand for fossil fuels and the amount of consumption are likely to increase.
- Government-owned energy companies and TNCs pay for extraction, refining and transportation because they can make huge amounts of money selling the fuel.
- Natural resources like water, wind and sunlight are known as renewable resources and can be harnessed to provide electricity.
- Renewables are not able to meet all current energy demands, but non-renewables can meet the demand for now.
- Renewables may not be suitable for all countries, e.g. solar power is unlikely to be suitable for countries that receive limited sunlight.

Key Point

Most resources are exploited by HICs, which also tend to consume most of these resources.

Extraction of mineral resources in a granite quarry

Power-generating equipment at a hydroelectric dam

Exploitation of Environments

- Humans exploit resources for economic gain in order to trade, for industrial purposes and to feed the 7 billion people living on the planet.
- Many **pristine** areas have been affected by our need for resources.
- Such areas include the tropical rainforest (for timber and metals), the oceans (for fish and fossil fuels) and the polar regions (for fossil fuels and precious minerals).
- In the tropical rainforests, this has led to problems such as deforestation, **biodiversity loss**, soil erosion, desertification, pollution of air and water and forced migration.
- Large areas of pristine forest have been removed to build mines, plant palm oil or soya bean plantations and introduce cattle ranches.
- These land-use changes have an impact on climate change both globally and locally.
- Humans have also harnessed the power of water in the tropical rainforests to build large **multipurpose dams**.
- In some polar areas such as Alaska, humans have exploited oil resources in areas like the northern coastal regions even though the nearest accessible port is over 1000 km away.
- A pipeline has been built across environmentally sensitive wildernesses to link the oilfield with the port of Valdez.
- The world's oceans have been overfished to the extent that some species have become extinct, while drilling for oil has led to pollution, with disastrous impact on the marine environment.

Key Point
The exploitation of natural resources by humans can lead to environmental problems.

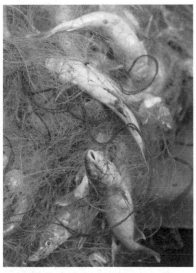

Overfishing threatens the existence of some species

Oil pipeline on the North Slope of Alaska

Quick Test

1. Suggest why the world's resources are not evenly distributed.
2. Why are natural resources highly prized by governments?
3. Why have humans been forced to exploit pristine and inaccessible places to obtain natural resources?

Key Words
natural resources
abiotic
biotic
fossil fuel
pristine
biodiversity loss
multipurpose dam

Food 1

You must be able to:

- Explain how global demand for food is rising but supply can be insecure, which may lead to conflict
- Describe the range of strategies that can be used to increase the supply of food.

Global Food Supply

- **Food security** is defined as 'having reliable access to a sufficient quantity of affordable, nutritious food' (World Health Organisation, WHO).
- In many developed countries, food security is the norm and the population are well fed and nourished.
- **Food insecurity** is defined as 'being without reliable access to a sufficient quantity of affordable, nutritious food' (WHO).
- More than 800 million people, mainly in developing countries throughout the world, live with hunger or food insecurity.
- Food production is influenced by two factors – the physical environment and human capacity – and is not evenly distributed throughout the world.
- Climate, water availability and soil type are the main components of the physical environment, while human capacity refers to both the population size and farming skills as well as the financial investment a country is able to put into its agriculture.
- Countries with vast populations like China and India have agricultural output values in excess of $100 billion.
- HICs such as the USA have very large output values due to the intensification of farming methods and large investments of capital.
- In contrast LICs, like those found in Sub-Saharan Africa, produce less food.

Poor physical conditions can contribute to food insecurity

 Key Point

HICs have high levels of food security, while LICs tend to have low levels.

Global Increase in Food Consumption

- Calorie intake varies between rich and poor countries.
- People in the USA and the European Union consume the most calories per capita on average (3000 per day).
- This is much higher than the recommended daily calorie intake of 2500 for men and 2000 for women, leading to obesity in these countries. A high calorie diet is also associated with cardiovascular disease and diabetes, as well as other health problems.
- Sub-Saharan Africa has many countries experiencing below average per capita calorie intake.

- Many LICs have rising populations and as the world's population increases by around 80 million people each year, more land is put under cultivation to feed them.
- As countries develop economically, their population demands a more western-style diet, with more meat and dairy products.
- This leads to grain, that was once used to feed the people, being fed to animals, and this results in shortages for the poor and diet-related illness for the rich.

Factors Affecting Food Supply and Food Security

- Extreme climatic conditions in some countries (such as drought or floods) can lead to low agricultural **yields** and environmental degradation (such as desertification), in turn leading to **famine**.
- Famine causes rising food prices. Other factors can also cause rises to food prices, such as natural disasters like drought, leading to under-nutrition and starvation among the poorest in those countries.
- **Climate change** has also had an impact in some of the countries of Africa, as soil erosion and desertification have led to a lack of land to grow food on.
- The countries that face the biggest food shortages, and therefore the worst levels of food insecurity, are often those that face the biggest threats from agricultural pests and diseases.
- Those countries also face the biggest risk from water shortages caused by unpredictable rainfall.
- The Sahel region of Africa has been suffering from drought on a regular basis since the early 1980s. The 1983–85 famine in northern Ethiopia was the worst to hit the country in a century and led to 400 000 deaths.
- The effects of the famine were made worse by a background of social unrest and civil war, which had raged in Ethiopia and Eritrea for two decades.
- The introduction of technology into agricultural methods in LICs can lead to massive growth in the amount of food being produced.

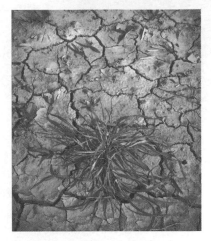

Poor soil conditions hinder food production

 Key Point

Producing food sustainably is a challenge for most nations.

Technology can help to increase food production

Quick Test

1. What is food security?
2. Suggest physical reasons for food insecurity.
3. Suggest human reasons for food insecurity.

 Key Words

food security
food insecurity
yield
famine

Food 2

You must be able to:

- Describe a range of strategies that increase food supply
- Explain the ways in which food production can be made more sustainable.

Strategies to Increase Food Supply

- **Irrigation** is the application of water to assist in the production of crops.
- Irrigation is used to increase the yields of crops, especially in those places where water is in short supply or in those places where the plentiful supply of water can be seasonal.
- The Gezira Scheme in Sudan is an example of a large-scale irrigation project, where the floodwaters of the Nile are used to grow crops such as cotton and wheat. However, large volumes of water are taken out for this process, which affects other irrigation schemes downstream.
- Aeroponics is a way of growing plants in an air or misty environment without the use of soil, e.g. NASA have experimented with producing fresh food for astronauts travelling to distant planets in the future.
- Hydroponics is a method of growing plants using nutrient-rich water, often using inputs of ultra-violet light for high-value crops like tomatoes grown in glasshouses.
- The Blue Revolution refers to the growth of aquaculture as a very highly productive way of producing food in both salt water and fresh water.
- Fish farming, such as shrimp or salmon farming, and the gathering of seaweed are all methods of aquaculture.
- The Green Revolution is an initiative that increased agricultural production, particularly in LICs, from the late 1960s.
- The Green Revolution saved around a billion people from starvation through the development of high-yielding varieties of cereals (rice and wheat) and new farming techniques, including the use of irrigation, synthetic fertilisers and pesticides.
- Biotechnology uses biological processes to develop new products that can help feed the hungry by producing higher crop yields.
- Biotechnology can lower the input of agricultural chemicals into crops, resulting in fewer vitamin and nutrient deficiencies and reduced allergens and toxins.

Irrigation

> ### Key Point
>
> Technology can be used to increase food supply and reduce food insecurity.

Towards a Sustainable Resource Future

- Organic farming is a sustainable method of food production that does not use synthetic pesticides or fertilisers. Organic products, such as plant and animal waste, are used as fertiliser and biological solutions control pests and disease.
- Urban farming is a sustainable practice where growers cultivate and process produce, often for their own use, and can include activities such as vegetable and fruit growing or beekeeping.
- The Organopónicos in Havana, Cuba, is a good example of a scheme to increase food supply for local, urban families.
- The scheme started in the 1990s when the Soviet Union collapsed and could no longer provide food to Cuba. It consists of low-level concrete walls filled with organic matter and soil, with lines of drip irrigation laid on the surface.
- Around 35 000 hectares of Organopónicos are found in Havana, growing various fruit and vegetables.
- **Ethical consumerism** is the purchase of sustainable sources of food, such as locally grown food and only buying fruit and vegetables that are in season. This helps to keep food miles to a minimum.
- Other examples of living sustainably include changing diets to those less reliant on meat and dairy products, and using sustainable fish sources such as pollock instead of cod (allowing the cod to recover in overfished areas).
- Reduced food waste and loss in both homes and in shops could save UK consumers up to £2.4 billion a year. The average UK family throws away enough food for around six meals a week.

An urban farming initiative in Detroit, USA

Throwing away food is a waste of money

Population and Food Supply

- The English 18th-century clergyman Thomas Malthus concluded that the rate of population growth was faster than the rate that food supplies could grow.
- Over time, population would outstrip food and some people would starve and the population would decline.
- In this theory, more people may die due to wars breaking out to secure food and others would die from malnourishment.
- Malthus suggested smaller families would help to resolve the problem.
- There have been many local famines, particularly in Africa, which have checked population growth in the way Malthus predicted.
- The Danish 20th-century economist Ester Boserup took a more positive view, writing 'necessity is the mother of invention'. This means someone will always invent a technological fix for any problem.
- The Green and Blue Revolutions and biotechnology are examples of Boserup's view.

Key Point

Sustainable use of food resources can also lead to increased food supply and reduced food insecurity.

Quick Test

1. How does technology increase crop yields?
2. How can changes to eating habits reduce our carbon footprint?
3. What is ethical consumerism?

Key Words

irrigation
ethical consumerism

Water 1

You must be able to:

- Explain why some areas have water surpluses and some have deficits
- Describe how and why water use varies in different countries.

Global Supply of Water and Impacts

- Fresh water is very rare – only 3% of the world's water is fresh, with two-thirds of that stored in ice caps and glaciers and unavailable for human use.
- The supply of water is uneven because the distribution of rainfall varies from place to place.
- Regions of the world that have water surplus have **water security**, while those with a deficit have **water insecurity**.
- For example, the UK, France and Germany have water security; Turkey, Israel and Jordan have water insecurity.
- Some regions have a **physical water scarcity** through natural reasons for low water supply, e.g. low rainfall in deserts.
- **Economic water scarcity** is low supply caused by human factors, such as the lack of a water infrastructure that can be found in many LICs.
- The wettest places in the world, such as rainforests, do not support many people, while the number of people living in **arid regions** is increasing and putting a greater demand on available resources.
- Low rainfall can cause other issues for water availability, such as low supplies of **groundwater**, an absence of lakes or rivers, and arid conditions meaning that accessing water is difficult.
- In many LICs, regions with limited water supplies lose up to 75% of the water used in irrigation through leakage, evaporation or run-off.
- Some 1.1 billion people worldwide lack access to safe drinking water and a total of 2.7 billion find water scarce for at least one month of the year.
- Poor sanitation is a problem for 2.4 billion people as they are exposed to diseases, such as cholera, typhoid and other **waterborne diseases**.
- Two million people, mostly children, die each year from diarrhoea.
- Many rivers, lakes and aquifers are drying up or becoming too polluted to use and more than half of the world's wetlands have disappeared.
- Agriculture consumes more water than any other user and wastes much of that through inefficient management.

Fresh water is not readily available across much of the world

 Key Point

Over 1 billion people do not have access to safe drinking water.

- Climate change is altering patterns of rainfall around the world, causing shortages and droughts in some areas and floods in others.
- By 2025, two-thirds of the world's population may face water shortages.

Global Demand for Water and Impacts

- As the world's population rises, so does the demand for water for drinking, bathing, agriculture and industry.
- With economic development, the domestic demand for water also rises through the use of more labour-saving devices, such as washing machines and dishwashers.
- The average North American uses 570 litres of water per day, while the average European uses 130 litres and the average African around 20 litres.
- With economic development, industrial and agricultural use of water also increases due to a greater demand for goods and services, such as meat and dairy products.
- Increasing water usage by humans has resulted in the depletion of groundwater supplies – a non-renewable resource – while rivers are often diverted for agricultural and industrial purposes.
- Studies in Africa and Asia show that the poorest 20% of the population spend up to 11% of their household income on water.
- The human right to safe drinking water was first recognised by the United Nations as part of international law in 2010.
- The right to water entitles everyone to have access to sufficient, safe, acceptable, physically accessible and affordable water for personal and domestic use.

Water being sprayed on a strawberry field

Key Point

Water use increases with rising levels of development and population growth.

Economic development brings with it a rise in demand for water

Key Point

Agriculture is the biggest user of water.

Key Words

physical water scarcity
economic water scarcity
arid region
groundwater
waterborne disease

Quick Test

1. Define 'water security'.
2. Where is most of the world's fresh water stored?
3. Why are water supplies in some places becoming scarce?

Water 2

You must be able to:

- Describe schemes to transfer water between regions with surpluses and those with deficits
- Explain strategies to obtain and conserve water sustainably.

Water Transfer Schemes

- In South Africa, a **water transfer scheme** is in operation where water is transferred from an area with a plentiful supply like the Lesotho Highlands to an area where water is in short supply, such as Gauteng province along the Vaal River.
- The Vaal River has a number of dams that supply agricultural areas such as the Vaalharts irrigation system (one of the most productive in the country), which covers about 44 000 hectares with a variety of crops.
- The Vaal basin serves the most populated and industrialised part of the country, including Johannesburg, and receives water from seven different transfer schemes including the giant Lesotho Highlands Water Project.
- Another major water transfer scheme is the South-North Water Transfer Project in China, which will enable 44.8 billion cubic metres of water per year to be transferred from the Yangtze River in the south to the Yellow River Basin in the arid north.
- As climate change makes droughts more common, a growing number of countries are turning to **desalination** – removing salt from seawater or groundwater to make it drinkable.
- Since 2010, London has had a seawater desalination plant to help address the problem of the South East being a water-stressed area, especially during dry years.

> ### Key Point
>
> In countries where there is a shortage of water, water is transferred between regions with surpluses to those with deficits.

Sustainable Management Strategies

- Water can be conserved through sustainable management strategies.
- **Water conservation** methods can take place in the home and can include low-flush toilets, taking short showers and only using washing machines and dishwashers when full.
- In the garden, methods can include growing drought-resistant plants that need little watering, spreading mulch or composted bark on flower borders and connecting a water butt to drainpipes and using the rainwater to water plants.
- Groundwater makes up nearly 30% of all the world's fresh water and is often used in places where there is not enough water to drink. It is often found in porous rocks deep underground in reservoirs called **aquifers**.

A water butt

- Groundwater quality is usually very good and needs less treatment than river water to make it safe to drink, as the rocks through which the groundwater flows help to remove pollutants.
- Groundwater also responds slowly to changes in rainfall, and so it stays available during the summer and during droughts, when rivers and streams have dried up.
- Groundwater is relied on in many developing countries, especially those in Africa where aquifers come close to the surface and the water can be accessed by shallow wells in remote villages.
- Extracting groundwater is relatively inexpensive as it is often drawn from wells – this does not require much technology and means that reservoirs are not needed to store the water.
- Aquifers are often slow to recharge (fill up) so may not always be sustainable.
- Small-scale sand dams are used in many arid countries as a way of storing precious water.
- They consist of a reinforced concrete wall built across a river bed that has seasonal flow. They are a simple, low cost and sustainable solution to the storage of water.
- Water used in the home can be recycled by treating it in wastewater (or sewage) plants.
- **Greywater** harvesting is a way of conserving water, involving the recycling of water used in baths and showers, as well as rainwater from roofs, to be used for flushing toilets and other non-drinking water purposes. Many modern buildings in the UK use greywater for these purposes.
- Local sustainable water schemes are found in many LICs; however, many are beset with problems, such as the Agra clean water project in northern India.
- This scheme is designed to provide water to the poorest areas of Agra, where waterborne diseases are a persistent problem owing to polluted sources.
- The solution has been partially solved by bottling clean drinking water from 130 km away, which is funded by UK charities and non-governmental organisations (NGOs).
- The scheme has many issues that limit its success, especially a lack of money for the locals to buy the water, little local technical know-how, and few spare parts for the machinery needed for the scheme to operate.

Water recycling at a sewage treatment plant

Key Point

Conservation of water is an important strategy to save precious resources.

Quick Test

1. What strategies can be used to obtain water in areas where it is in short supply?
2. What methods of sustainable water management can be used in the home?
3. What are the advantages and disadvantages of using groundwater supplies?

Key Words

desalination
water conservation
aquifer
greywater

Energy 1

You must be able to:

- Explain how some areas of the world experience energy security while others have energy insecurity or even energy poverty
- Describe how exploiting energy resources has many impacts on supply and the environment.

Global Supply and Demand for Energy Resources

- Demand for energy resources is rising globally but supply can be insecure, which can lead to conflict.
- Much of the world relies on fossil fuels that are non-renewable; reducing our reliance on these will increase our **energy security**.
- Some countries and regions are energy secure; others suffer from **energy insecurity**.
- Eastern Europe, including Russia, has large reserves of natural gas and coal, with Russia among the top ten countries for oil and uranium production and thus energy secure.
- Much of the rest of Europe is heavily dependent on energy imports as it has declining fossil fuel supply and has energy insecurity.
- The Middle East and North Africa have large oil reserves but unstable governments often affect how much oil they can produce, though at present they are energy secure.
- North America has vast coal resources but its conventional oil resources are largely now exploited. It is now exploiting non-conventional oil and gas reserves in Alaska, the Alberta Tar Sands and through fracking, but its huge **energy consumption** often outweighs supplies and means it is still, to some extent, energy insecure.
- Most of Asia (excluding Russia) has large coal and uranium reserves but has a rapidly increasing demand for oil in particular, outweighing available supplies. The region suffers from energy insecurity.
- Sub-Saharan Africa depends on foreign TNCs to exploit resources, e.g. oil in Nigeria and Sudan.
- Many African people use wood as their primary source of energy and much of the population suffers from **energy poverty**.
- Energy consumption is increasing throughout the world for a number of reasons:
  - Economic development leads to increased consumption.
  - As rising wealth leads to increased living standards, this in turn leads to more electrical items and technology being used in the home and increased rates of car ownership.
  - As the world's population rises, so does the demand for energy.

There are an estimated 892 billion tonnes of proven coal reserves worldwide – enough to last about 100 years at current rates of production

 Key Point

Securing supplies of reliable energy at an affordable price is challenging for many countries.

Factors Affecting Energy Supply

- Physical factors such as the following all have impacts on the availability of energy:
  - geology
  - the remoteness of the energy sources
  - climate.
- Fossil fuels are found in sedimentary strata of varying ages that may be found in hard-to-reach locations, and that impacts on the cost of production.
- Climatic factors lead to high energy demands at different times of the year.
- Climatic factors also lead to other problems, such as:
  - cloud cover affecting the availability of solar energy
  - wind speed affecting the availability of wind energy.
- The development of technology has enabled humans to exploit energy sources in remote, difficult and environmentally sensitive areas, such as in deep oceans and polar regions.
- However, the variation in fossil fuel prices, especially oil and gas, can mean that the cost of exploitation and production in areas that are challenging becomes prohibitive, leading to shortages in the supply of these fuels.
- Political factors also affect the supply of energy, such as the conflicts in Middle Eastern countries in the late 20th and early 21st centuries (e.g. Iraq and Syria are heavily involved in the control of oilfields).

A former drilling tower in the Arctic

Oil wells on fire outside Kuwait City in the aftermath of the First Gulf War, 1991

> **Key Point**
>
> Relying less on fossil fuel will improve energy security.

Quick Test

1. What regions of the world are energy secure and why?
2. Suggest how energy consumption rates vary in different climates.

> **Key Words**
>
> energy security
> energy insecurity
> energy consumption
> energy poverty

Energy 2

You must be able to:

- Explore how exploitation of new energy resources often takes place in environmentally sensitive and challenging places
- Describe how energy prices impact on many areas of our lives
- Explain how renewable energy is making significant advances in the energy mix of the world.

Exploitation of Energy Resources and Impacts

- The rising demand for gas and oil has led to energy TNCs exploring and prospecting for oil in geographically remote, difficult and environmentally sensitive areas.
- In the 1960s, British companies searched for oil and gas in rocks underneath the North Sea, leading to the UK becoming self-sufficient in oil and gas for a while.
- In Alaska, the exploitation of oil sources has taken place in environmentally sensitive and remote areas close to the Arctic Ocean.
- A controversial 1200 km long pipeline was built at a cost of $8 billion to link the north of Alaska and the port of Valdez in the south.
- Agriculture uses oil products to power farm machinery and to transport goods from the farm.
- Oil is used in agricultural chemicals such as fertilisers and pesticides, and if the oil price increases so will the price of food.
- In recent years, agricultural goods like barley, maize and sugar cane have been used to make **biofuels**.
- Industry also relies on oil to power machinery and transport goods around.
- Oil is also used as a raw material in many industries, such as plastics, packaging and textiles.
- Increases in the oil price lead to higher prices for goods.
- Oil has also led to many conflicts, especially when the control of oil resources was at stake such as during unrest in the Middle East.
- In Canada, oil is being extracted from sand in the province of Alberta. This has been highly controversial and has caused huge amounts of pollution, resulting in illness in the local population and extensive environmental damage.
- The Keystone XL pipeline links these oilfields with refineries in the USA and again has caused controversy as extensions to the pipeline were cancelled by then-president Barack Obama but later reinstated by his successor, Donald Trump.

Key Point

Humans are having to exploit extreme environments in order to find new oil reserves.

Key Point

Oil price rises impact on the world's economy, including agriculture and industry, as well as transport.

Strategies to Increase Energy Supply from Renewable Sources

- Renewable sources of energy are more environmentally friendly than non-renewable sources but they can be more expensive to generate electricity from as the set-up costs can be greater.
- **Biomass** is the use of plants to produce energy; this includes burning wood and converting crops into biofuels such as ethanol.
- Many power stations in the UK have been converted to burn biomass as well as coal. Drax power station in South Yorkshire, originally built to burn coal from the nearby Selby coalfield (now closed), now burns coal, wood pellets and processed straw.
- Wind power is the most widely used renewable in the UK with nearly 7000 onshore and offshore wind turbines producing almost 10% of the country's electricity and making the UK the world's sixth-largest producer of wind power.
- The Muppandal wind farm in Tamil Nadu in southern India has over 3000 wind turbines and is one of the largest in the world. It produces 1500 MW – about 20% of India's total wind power.
- **Hydroelectric power** (HEP) is the most established of the renewables as massive dams in many parts of the world produce vast amounts of electricity, such as the Three Gorges Dam in China.
- In many remote areas, micro-hydro schemes generate electricity in places that may not be connected to electricity networks.
- An example is Chambamontera in the Andes mountains of Peru, where a non-governmental organisation, Practical Action, has supplied equipment to a remote community to produce enough electricity for their needs.
- Tidal and wave power are forms of HEP that use the sea to produce electricity. Most tidal schemes use barrages that trap the incoming tide and release it when electricity demand is high to drive turbines and produce power.
- A tidal power station has been operational since 2007 in Strangford Lough in Northern Ireland.
- **Geothermal energy** is found mainly in areas where there is volcanic activity (such as Iceland) and the steam produced is used to power turbines to create electricity.
- Solar power has massive potential in countries with high annual amounts of sunshine. Even in the UK, with over 650 000 solar power installations, up to 3% of total electricity generation is possible if the conditions are favourable.

Wind turbines and solar panels

A geothermal power plant

Quick Test

1. What type of challenges do companies face in drilling for oil in places like Alaska?
2. What is the most important renewable source of energy for the UK? Why?

Key Words

biofuel
biomass
hydroelectric power
geothermal energy

Energy 3

You must be able to:

- Describe the strategies to increase energy supply through the use of non-renewables
- Explain approaches to energy conservation in the home, work and transport.

Strategies to Increase Energy Supply from Non-Renewable Sources

- Fossil fuels are natural fuels such as coal, oil and gas, formed in the geological past from the remains of living things.
- All fossil fuels produce carbon dioxide (CO_2) that contributes to the **enhanced greenhouse effect** which causes climate change.
- Most countries in the world are committed to cutting their CO_2 emissions from fossil fuels.
- Many countries, including the UK and Germany, have used gas to replace coal as a source of energy to make electricity, because gas produces only half as much CO_2 as coal.
- **Clean coal technology** has enabled coal-fired power stations to produce less pollution and CO_2 by removing the pollutants when the coal is burned. However, in the UK they will all close by 2025.
- Nuclear power is seen by many to be the solution to electricity generation as it is a carbon-free source of energy (apart from in the construction phase and the transport of fuels).
- Some countries, such as the UK, are looking to increase their number of nuclear power stations to maintain energy supply while reducing greenhouse gas emissions.
- New efficient nuclear power stations like the proposed Hinkley Point C in Somerset could produce 10% of the UK's electricity.
- A network of around five to six nuclear power stations could produce over half of the UK's electricity needs in the future.
- Some countries already have large numbers of nuclear reactors that produce much of their electrical energy. For example, France has 58 reactors producing 76% of its electricity.

Cooling towers of a nuclear power plant

Key Point

Fossil fuels produce carbon dioxide and many countries are looking for low carbon alternatives.

Energy Conservation

- A **carbon footprint** is the greenhouse gas emissions produced by an organisation, event, product or individual.
- Carbon footprints in HICs tend to be high because of the lifestyles people lead, with a reliance on fossil fuels, heavy use of electrical devices at work and home, and a diet that relies on food that is imported or uses high levels of inputs derived from fossil fuels.
- In LICs, carbon footprints tend to be low because lifestyles are more sustainable as less fossil fuel is used and food tends to be locally produced.

Energy-efficient light bulbs are a simple energy conservation method

- **Energy conservation** is important in protecting precious supplies of energy.
- Energy can be conserved through the intelligent design and construction of homes and other buildings like offices and factories. Methods include the use of insulation in loft spaces and walls, double-glazed and triple-glazed windows, and larger windows in south-facing walls.
- Using energy-efficient devices, like kettles that only boil one cup of water, or turning off appliances when not in use (rather than using standby buttons) can also help conserve energy in the home.
- The use of LED light bulbs, washing machines that wash clothes at 30°C or lower and thermostats on central heating systems that can be controlled using a mobile phone, are all ways in which homes and workplaces can be made more energy efficient.
- Transport in cities can be made more sustainable by encouraging people not to use cars. Public transport or bicycles are more sustainable and help to reduce carbon emissions.
- **Hybrid** buses have been introduced in many cities like London, Reading and Brighton, and they combine a conventional engine with an electric one, which results in greater energy efficiency and lower emissions.
- In many UK cities, the provision of bikes to hire at stations and bus stops has been popular following the example of London, which introduced so-called 'Boris Bikes' (named in 2010 after the mayor at the time, Boris Johnson). There are now more than 11 000 of these bikes covering the central part of London, and other cities like Liverpool have followed suit by introducing bikes for hire.
- Cars have become more energy efficient, with the introduction of new engines that burn fuel more efficiently, hybrid engines in cars like the Toyota Prius and more aerodynamic designs.
- The UK uses about the same amount of oil for transport as it did in the 1970s despite the number of vehicles doubling since then.
- Airliners too have seen similar levels of efficiency with the introduction of aircraft like the Boeing 787 Dreamliner. The Dreamliner uses 25% less fuel than conventional airliners like the Airbus 340 due to efficient engines and the use of lighter materials in the aircraft's construction.
- Education is important in making the public aware of energy conservation. Examples of energy conservation promotion include the use of posters beside light switches and the inclusion of the topic in GCSE courses.

Some cities are making the streets more bike-friendly

Dreamliner aircraft

Key Point

Strategies to use energy more efficiently will lower demand.

Quick Test

1. Why is the use of fossil fuels unsustainable?
2. Why might generating electricity from nuclear power be a better option than the continued use of fossil fuels?
3. Suggest reasons why carbon footprints are greater in HICs than LICs.
4. How can we conserve energy in our homes?

Key Words

enhanced greenhouse effect
clean coal technology
carbon footprint
energy conservation
hybrid

Fieldwork

You must be able to:

- Answer questions in an examination showing that you have made links between two fieldwork enquiries about the physical and human geography you have studied.

What is the Geographical Enquiry?

- A **geographical enquiry** is an investigation linking your study in class with fieldwork carried out away from school.
- You will have completed two geographical enquiries in two different places, studying the physical and human geography at these locations.

What Sort of Questions Can I Expect in the Examination?

Questions About How You Might Use Your Fieldwork Techniques in Unfamiliar Contexts

- You may be presented with new information and data about an investigation.
- What techniques were used or would you suggest using for collecting such data?
- What methods were used or would you suggest for presenting the data?
- How effective would these techniques be?

Students measuring orientation of glacial deposits in Snowdonia

Questions About Your Own Enquiries

- You will need to know the titles and be able to describe the location for both of your enquiries.
- Why did you choose the location for your enquiries?
  - Explain why the locations were suitable.
- Why did you choose your question?
  - How is it linked to geographical theories or concepts?
  - Explain any **hypothesis** or hypotheses you tested. A hypothesis is an idea or explanation for something that has not yet been proved. It could be described as an 'educated guess' about the possible outcome of your investigation.
- How did you prepare for your fieldwork?
  - In particular, what risk assessment did you conduct (an analysis of possible dangers) for both enquiries and what measures did you take to reduce these risks?

Field sketching in Carding Mill Valley, Shropshire

- Why did you collect your data?
  - You will need to describe and explain the **primary data** (data you and your fellow students have collected) and the **secondary data** (data collected by another person, group or organisation that you have accessed via articles, websites, books, etc.).
- Is your data **quantitative** (statistics; numbers)?
  - For example, width and depth of a river channel; an environmental quality survey.
- Or is your data **qualitative** (non-numerical; might involve subjective judgements)?
  - For example, field sketches; photographs; video; quotes or opinions.
- How did you collect your data?
- What methods of data collection did you use?
  - Mention **sampling techniques** you used to ensure the reliability of your data, e.g. random; systematic; stratified. If you used **GIS**, explain why such georeferenced data is useful.
  - Justify a method of data collection that you used.
- How did you process and present your data?
- Evaluate a method of data presentation that you used.
  - Describe and explain what you did with your data to make sense of it; what visual, graphical or cartographic methods (including GIS) you used and why.
- What did your data tell you?
  - Describe, analyse and explain your data.
  - What links did you find between data sets?
  - Which statistical techniques did you use?
  - What patterns and anomalies did you find?
- What conclusions did you make?
- Did your conclusions match your expectations at the beginning of the enquiry?
  - Make sure you refer to the original aims or hypotheses. A hypothesis may be correct or incorrect – it is important that you can explain why.
- How effective was your investigation?
  - How could you improve your data collection techniques?
  - What were the **limitations** of your data?
  - How reliable were your conclusions?

A student measuring river channel gradient in Carding Mill Valley, Shropshire

> **Key Point**
>
> For both of your enquiries, you need to be able to describe and explain what you did at each stage and why. You must also be able to reach plausible conclusions supported by your data, evaluate your work and recognise any limitations.

Quick Test

1. What is a hypothesis?
2. What is the difference between primary and secondary data?
3. What is the difference between quantitative and qualitative data?

> **Key Words**
>
> geographical enquiry
> hypothesis
> primary data
> secondary data
> quantitative data
> qualitative data
> sampling techniques
> GIS
> limitations

Issue Evaluation

You must be able to:

- Apply geographical knowledge, understanding and skills to a particular issue.

What is the Issue Evaluation?

- The issue evaluation encourages **synoptic thinking**, which means applying the knowledge, understanding and skills from all of your geographical learning to a new situation at a range of scales from local, regional, national to international.
- The topic for the issue evaluation will be linked mainly to one of the compulsory (core) topics, but there will also be links to other topics.
- 12 weeks before the exam, a 'pre-release' resource booklet will be published about the issue.
- You can make notes on your resource booklet, but you cannot take this copy into the exam – a new copy will be provided for candidates.
- You will need to show that you understand the issue by referring to the resources.
- You will need to demonstrate **critical thinking** about the issue, which means showing your ability to interpret, analyse and evaluate ideas and arguments.
- **Problem-solving** is required: you will need to make a decision by choosing a possible option and justifying your decision.
- Longer, extended written answers will be expected for some questions.

> ### Key Point
>
> The issue evaluation encourages synoptic thinking. You will have to draw upon knowledge from all of the geography you have studied.

What Geographical Evidence is in the Resource Booklet?

- The pre-release booklet will contain a range of resources, such as different kinds of maps of different scales, GIS data, charts, diagrams, tables of statistics, photographs, aerial or satellite imagery, field sketches and written materials (e.g. web or newspaper articles; quotes from **stakeholders**, **key players** and **interest groups**).
- All the resources are important. You will need to use evidence from them to demonstrate your understanding of the issue and to support the arguments for or against a point of view. It may well be useful to use more than one resource at a time.

> ### Key Point
>
> You will need to take into account a wide range of facts and views to inform your decision-making.

How Can I Reach a Decision?

- Consider these questions:
  - What are the **impacts**? Remember, impacts can be positive as well as negative.
  - What are the main sources of conflict and agreement, e.g. between stakeholders, key players and interest groups.
  - How sustainable is each option or strategy?
  - How can I structure my thinking? Use categories such as advantages versus disadvantages; costs and benefits; SWOT analysis (strengths, weaknesses, opportunities, threats); economic, social and environmental considerations (remember that there are different types of pollution – air, water, noise, light, litter, etc.).
  - Why is **management** important? There are different approaches to governance, e.g. top-down (i.e. decisions are made from the top, such as national government, usually for large-scale schemes) versus bottom-up (i.e. decisions are based on everyone's opinion, such as those of local people, usually for small-scale schemes). Sustainable decisions consult people so they feel empowered (i.e. they feel they have greater ownership of the project and are therefore prepared to invest time and energy in supporting it).
  - How should I reach a decision? Explain reasons for choosing one option and reasons for not choosing other options. There will not be a 'right' or 'wrong' answer, but whatever decision you make must be supported by evidence from the resources.

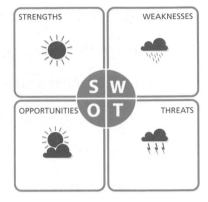

Key Point

There is no 'correct' or 'incorrect' answer, but your decision must be supported by evidence in the resources.

Quick Test

1. What is meant by 'synoptic thinking' and 'critical thinking'?
2. Give definitions, with examples, of:
   a) stakeholders
   b) key players
   c) interest groups.
3. Explain what is meant by 'top-down' and 'bottom-up' approaches to management.
4. What is meant by 'empowerment'?

Key Words

synoptic thinking
critical thinking
problem-solving
stakeholders
key players
interest groups
impact
management

Review Questions

Urbanisation

1 What is urbanisation? [1]

2 Identify two reasons why LIC cities are growing so fast. [2]

3 Which of the following statements are true? Which are false?

a) In 1900, 40% of the world's population lived in urban areas.

b) A mega-city has a population of over 10 million people.

c) Low wages are a 'pull factor' of rural to urban migration.

d) The majority of the world's population now live in urban areas.

e) The fastest rates of urban growth are experienced in higher income countries. [5]

4 Read each of these and decide whether they refer to either **push factors** or **pull factors** of rural to urban migration:

a) Unemployment

b) Higher wages

c) Isolation

d) Unprofitable farming

e) Better schools and hospitals [5]

5 What is a 'greenbelt' and how can it help to control urban sprawl? [2]

Total Marks _____ / 15

Urban Issues and Challenges 1

1 Describe in detail two ways that quality of life can be improved in favelas such as Rocinha. [4]

2 Identify three positive aspects of life in Rio de Janeiro's favelas. [3]

3 Describe one environmental challenge faced by Rio de Janeiro. [2]

4 Describe the living conditions in Rio's favelas. [4]

Total Marks _____ / 13

Urban Issues and Challenges 2

1 Describe two environmental challenges faced by London. [4]

2 Describe two transport developments that have taken place within London's Docklands. [4]

3 Give three reasons why tourists are attracted to London. [3]

4 Identify three social challenges faced by London. [3]

Total Marks .... / 14

Urban Issues and Challenges 3

1 Choosing from the list below, what percentage of the UK's population lived in urban areas in these years?

5%   17%   55%   77%   87%

a) 1700 [1]

b) 1900 [1]

c) 2010 [1]

2 a) What was the main reason that encouraged more and more people to move into UK cities from the beginning of the 1800s? [1]

b) Name two of the reasons why people were driven away from the countryside from the beginning of the 1800s. [2]

3 a) What are the main differences between the urban models devised by Burgess and Hoyt? [2]

b) Describe the nature of the land use surrounding a central business district in the Burgess model. [2]

4 Describe two environmental improvements to urban areas that can form features of sustainable living. [4]

Total Marks .... / 14

Review Questions

UK Population and Economic Change 1

1 Name one area of the UK that supports a low population density. [1]

2 What advantages are there of an ageing population? [2]

3 What is 'counter-urbanisation' and how can it contribute to problems in rural areas? [3]

4 Explain how social and economic changes can result from rural depopulation. [6]

Total Marks _____ / 12

UK Population and Economic Change 2

1 What is a 'honeypot'? [1]

2 With reference to examples, explain what problems are faced at 'honeypots'. [4]

3 Describe some ways that national parks attempt to manage high visitor numbers at certain locations. [4]

4 Employment levels in the secondary sector have declined since the 1980s. Using named examples, explain why this has been the case. [6]

Total Marks _____ / 15

Global Development 1

1 What components make up the Human Development Index (HDI) and why has it become popular as a measure of world development? [3]

2 Explain how an NEE (newly emerging economy) differs from an LIC (lower income country). [2]

3 What historic causes have contributed to global inequalities? [2]

4 Identify three population characteristics of LICs. [3]

Total Marks _____ / 10

Global Development 2

1 How can intermediate technology help to reduce the development gap? [4]

2 What factors attract many transnational corporations (TNCs) to locate in LICs? [3]

3 What does the term 'globalisation' mean? Use examples to illustrate your answer. [3]

4 What is 'tied aid'? [2]

Total Marks _____ / 12

Practice Questions

Overview of Resources – UK

1 What are the solutions to end the UK's reliance on fossil fuels? [2]

2 Why is agriculture important to the UK economy? [3]

3 Why is water use increasing in HICs like the UK? [3]

4 What strategies can be employed to supply water to regions like south-east England, where demand is greater than supply? [4]

Total Marks _____ / 12

Overview of Global Inequalities

1 Which regions of the world use most resources per person? [3]

2 Give reasons why the regions in question 1 use most resources. [2]

3 Why do the poorest regions on Earth consume fewer resources than the richest regions? [3]

4 Using an area you have studied, explain how humans have exploited a pristine wilderness and describe some of the problems caused. [5]

Total Marks _____ / 13

Food 1

1 Suggest two different physical causes of food insecurity. [2]

2 Suggest two different human causes of food insecurity. [2]

3 Describe the differences in calorie intake between HICs and LICs. What are the health risks associated with a high calorie diet? [4]

4 Describe the potential consequences of climate change on food production for the poorest LICs. [4]

Total Marks _____ / 12

Food 2

1 What is meant by the term 'irrigation'? [1]

2 What is meant by the 'Blue Revolution'? [2]

3 Explain how any **two** of the following are able to increase food supply:

hydroponics   **aeroponics**   **biotechnology**   **appropriate technology** [4]

4 For a large-scale irrigation scheme you have studied, describe how its development has both advantages and disadvantages. [5]

> **Total Marks** .... / 12

Water 1

1 Define the term 'water insecurity'. [2]

2 Suggest two causes of water insecurity. [2]

3 Why will climate change have an impact on water security for many countries? [3]

4 As countries become more economically developed, their use of water increases.

Explain why this happens. [4]

> **Total Marks** .... / 11

Water 2

1 What is meant by a 'water transfer scheme'? [1]

2 What is meant by 'desalination'? [1]

3 Explain why the use of groundwater from aquifers is unsustainable. [3]

4 Describe the main features of a water transfer scheme you have studied. [5]

> **Total Marks** .... / 10

Practice Questions

Energy 1

1. Define 'energy security'. [1]

2. Define 'energy poverty'. [1]

3. Suggest two causes of energy insecurity. [2]

4. Why are regions like North America and western Europe energy insecure? [4]

Total Marks _____ / 8

Energy 2

1. How can water produce electricity? [2]

2. Why might solar energy not be a good option for the UK in order to generate large amounts of electricity? [3]

3. Give two advantages and two disadvantages of wind power for an HIC such as the UK. [4]

4. Explain why the rising price of oil leads to higher food prices. [5]

Total Marks _____ / 14

Energy 3

1 What are 'fossil fuels'? [1]

2 What is the 'enhanced greenhouse effect'? [2]

3 Why are the owners of coal-powered electricity generating stations having to introduce new technology, such as clean coal technology? [3]

4 Complete the table to show the advantages and disadvantages of nuclear power.

Advantages	Disadvantages

[8]

Total Marks _____ / 14

Fieldwork

1 Suggest two data collection techniques that could be used to carry out a geographical fieldwork investigation in:

a) a physical environment [2]

b) a human environment. [2]

2 Explain some advantages of the locations for each of your fieldwork enquiries. [2]

3 Justify one primary data collection method used in relation to the aim(s) of your physical/human geography enquiry. [3]

4 Explain how a data collection technique used in one of your enquiries could be improved to make the sample more reliable. [3]

Total Marks _____ / 12

Review Questions

Overview of Resources – UK

1. What is meant by 'organic produce'? [2]

2. Give two reasons for the development of fracking (hydraulic fracturing) in the UK. [2]

3. Why does the UK have energy insecurity? [2]

4. Why is the development of fracking likely to cause conflicts between different groups of people? [3]

Total Marks _____ / 9

Overview of Global Inequalities

1. Define the term 'natural resources'. [1]

2. What is meant by 'biodiversity loss'? [2]

3. Suggest reasons why use of the world's resources is likely to increase in the future. [3]

4. Giving an example, why do humans exploit areas of pristine wilderness? [4]

Total Marks _____ / 10

Food 1

1. What is meant by 'famine'? [1]

2. Explain how LICs can improve their food security. [3]

3. List some of the physical and human factors that affect food production. [4]

4. Why do some parts of the African continent have food shortages? [4]

Total Marks _____ / 12

Food 2

1 How does ethical consumerism make agriculture more sustainable, especially in HICs? [2]

2 How can reducing food waste in the home enable us to be more sustainable? [2]

3 Describe the main points of Malthus's and Boserup's theories. [4]

4 Describe the main features of a scheme in an LIC or NEE to increase sustainable supplies of food. [5]

> **Total Marks** _____ / 13

Water 1

1 Explain the term 'physical water scarcity'. [1]

2 Explain the term 'economic water scarcity'. [2]

3 Explain how LICs can improve their water security. [2]

4 Why do more than one billion people worldwide lack access to safe drinking water and what kind of health risks arise as a result? [4]

> **Total Marks** _____ / 9

Water 2

1 Suggest ways that water supplies can be increased in areas where water is scarce. [3]

2 Suggest ways in which water resources can be conserved in HICs. [3]

3 Why is groundwater an important water source for many countries? [5]

4 Describe the main features of a scheme that you have studied in an LIC or NEE to increase sustainable supplies of water. [5]

> **Total Marks** _____ / 16

Review Questions

Energy 1

1 Define 'energy consumption'. [1]

2 Study the pie charts below. They show the sources of energy used to make electricity in the UK in 1970 and 2014.

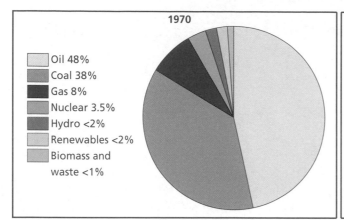

1970

Oil 48%
Coal 38%
Gas 8%
Nuclear 3.5%
Hydro <2%
Renewables <2%
Biomass and waste <1%

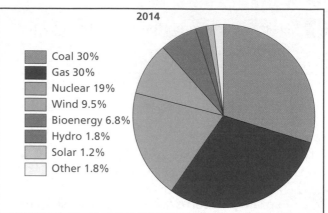

2014

Coal 30%
Gas 30%
Nuclear 19%
Wind 9.5%
Bioenergy 6.8%
Hydro 1.8%
Solar 1.2%
Other 1.8%

a) Describe the changes in the use of fossil fuels. [3]

b) What percentage of energy was made up of renewables in 2014? [1]

c) Why has the percentage of energy made by renewables increased since 1970? [3]

Total Marks .... / 8

Energy 2

1 What other uses are there for oil apart from as a fuel? [2]

2 Why might the growing of arable crops to make biofuels, such as biodiesel, be a controversial issue? [2]

3 Why are some countries not suited for electricity production from hydroelectric power? [3]

4 Describe the main features of a renewable energy scheme that you have studied in an LIC or NEE to provide sustainable supplies of energy. [5]

Total Marks .... / 12

Energy 3

1 Define 'carbon footprint'. [2]

2 Suggest reasons why carbon footprints are larger in HICs than in LICs. [4]

3 Describe how the design of buildings can be used to conserve energy. [3]

4 Why is the use of public transport or bicycles more sustainable than using cars in urban areas? [3]

> Total Marks .... / 12

Fieldwork

1 Describe a data presentation technique you used for your primary or secondary data. [3]

2 Explain why you chose a particular data presentation technique for your fieldwork enquiry, outlining the advantages and disadvantages. [4]

3 From the photograph, identify the potential hazards and who may be affected by them. [4]

4 Outlining some of your risk assessment, how would you reduce the risks posed by the hazards in your fieldwork? [4]

> Total Marks .... / 15

Mixed Questions

1 Study the figure below showing water usage per person in selected countries.

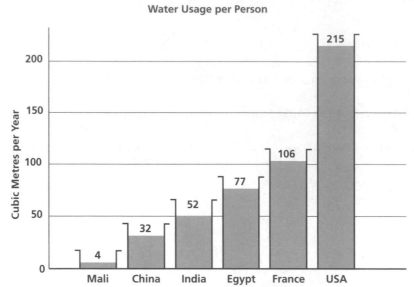

Give reasons for the pattern of water usage shown in the graph. [5]

....

....

....

....

....

2 Describe the distribution of arable and pastoral farming in the UK. [2]

....

....

3 What is 'intermediate technology'? Describe an example to explain how it can improve
people's lives. [5]

....

....

....

....

....

4 What is a 'dormitory village' and why do they develop? [3]

5 In what ways can the UK climate be influenced by continental air? [2]

6 What is the effect on mass movement of a long period of heavy rain? [4]

7 Use an example to describe one way in which an ecosystem naturally manages itself. [2]

8 What word is often used to describe northern coniferous *boreal* forests? [1]

9 What would a river in a previously glaciated valley look like today? [2]

10 What are the causes and potential impacts of a storm surge with a tropical storm? [4]

Mixed Questions

11 State two examples of secondary impacts of an earthquake.   [2]

12 How can a family use energy more efficiently in the home?   [4]

13 Many local residents opposed the redevelopment plans for London's docks. What might
    have been the reasons for this?   [3]

14 The UK has a temperate climate. What does 'temperate' mean?   [2]

15 Describe three ways that central business districts (CBDs) can be improved to encourage
    people to live there.   [3]

16 How do deciduous trees protect themselves from winter frosts?   [1]

17 Look at this photograph taken in Newlyn, Cornwall.

a) What two forms of coast protection are shown in the photograph? [2]

b) Describe one negative issue for each form of protection. [4]

18 What is the 'rural–urban fringe' and what land uses are found there? [3]

Mixed Questions

19 Giving examples, suggest reasons why oil and gas resources are being exploited in remote, difficult and environmentally sensitive areas. [5]

20 Approximately what percentage of the UK's natural heathland remains intact? [1]

21 What happens to the soil in a tropical rainforest after vegetation cover is removed? [1]

22 Why is Antarctica referred to as a desert? [1]

23 The map below shows the Brandt Line of world development. Why do you think this is no longer used? [2]

In terms of development, the world can be divided into two halves

24 Using diagrams, suggest two building techniques that could reduce the impact of an earthquake. [4]

25 Which of the following is the main variable used by the Saffir–Simpson scale to measure the intensity of a tropical storm? Tick the correct answer. [1]

A   Duration of the storm in hours ☐

B   Width as measured from space ☐

C   Rainfall amount ☐

D   Wind speed ☐

26 Name one hydroelectric power scheme in a hot desert. [1]

27 Name three examples of companies operating in the 'e-tailing' industry. [3]

28 Describe some of the conflicts that exist between farmers and tourists within UK national parks. [4]

Mixed Questions

29 Describe the main features of a named and located waterfall in the UK that you have studied. [4]

....

....

....

....

30 What is the name of:

a) the machine that measures shaking in an earthquake? [1]

....

b) the recording sheet (shown below) produced by this machine? [1]

....

31 Describe how the 'Green Revolution' has been able to increase food production in lower income countries. [4]

....

....

....

....

32 Why are many Middle Eastern and North African countries energy secure? [4]

....

....

....

....

33 On a separate sheet of paper, describe the formation of meanders using only labelled and/or annotated diagrams. [4]

34 For one of your geography enquiries, to what extent were the results of this enquiry helpful in reaching a reliable conclusion(s)? [9 + 3 SPaG]

Total Marks _____ / 107

Answers

Page 9 Quick Test

1. Destructive boundary – two plates travel towards each other and collide, with the denser plate sinking below the other; Constructive boundary – two plates pull apart to create new land.
2. **Any two from:** They have always lived there; jobs; confidence in government to 'fix' things; 'it won't happen to me' attitude; fertile soils; valuable minerals; geothermal energy; tourism.

Page 11 Quick Test

1. Primary effects occur immediately while secondary effects occur later on, bringing more problems to those affected.
2. The focus is the point underground where the earthquake originates, while the epicentre is the point on the surface directly above the focus.

Page 13 Quick Test

1. Shield volcano
2. Pyroclastic flow – torrent of hot ash, rocks, gases and steam, moving at up to 450 mph;
   Lahar – 60 mph mudslide of melted snow and volcanic ash.

Page 15 Quick Test

1. Intense, low-pressure systems
2. Blocking high pressure (blocking anticyclone)
3. Differences in pressure. Coriolis force impact differs between Northern and Southern Hemisphere.
4. If a tropical storm becomes cyclonic, air around the centre spins as a vortex which creates an eye 30–65 km wide.
5. It loses its energy supply (warm water) and slows owing to friction (especially if the land is hilly).

Page 17 Quick Test

1. a) **Typhoon Haiyan**: over 6300 (most due to storm surge);
   Hurricane Sandy: 233
   b) **Typhoon Haiyan**: a Category 5 tropical storm; **Hurricane Sandy**: a Category 2–3 tropical storm.
   c) **Typhoon Haiyan:**
   Primary: widespread devastation of homes and businesses; landslides; storm surges: a storm shelter was submerged and many drowned.
   Secondary: infrastructure damage impeded relief efforts; economic: businesses closed, development halted; social: homelessness (1.9 million), disease, schools closed; environmental: ecosystems, farmland damaged.
   Hurricane Sandy:
   Primary: wind damage to infrastructure; floods due to heavy rain and storm surges; subway tunnels were flooded; flights cancelled; over 6 million customers lost power supplies; 7000 people in emergency shelters; New York Stock Exchange closed for two days.
   Secondary: economic: second costliest storm ever in USA (over $71.4 billion); social: homelessness (650 000); political: may have had an impact on presidential election; military: led to a review of climate change impact on national security; environmental: untreated sewage released into sea.
   d) **Typhoon Haiyan**: warm oceans in the western Pacific; climate change may have contributed. **Hurricane Sandy**: warm surface water in the Caribbean and Atlantic; climate change may have contributed.
   e) **Typhoon Haiyan:**
   Immediate: much of the Philippines under a state emergency; worldwide relief effort totalled over $500 million.
   Long-term: 3 Ps reviewed, authorities adopted proactive strategies; resilience was improved by tackling inertia.
   Hurricane Sandy:
   Immediate: Federal Emergency Management Agency coordinated 3 Ps; many eastern states declared states of emergency and widespread evacuations arranged; public services (e.g. schools) closed; President Obama arranged federal financial assistance.

Long-term: as the richest country in the world, the USA was well-resourced and the key players well-trained in implementing the 3 Ps, creating a high level of resilience; lessons had been learned from Hurricane Katrina in 2005.

Page 19 Quick Test

1. The prevailing winds are westerlies from the Atlantic Ocean, and the relatively warm, moist air from the North Atlantic Drift (or Gulf Stream) creates the maritime effect.
2. a) **Any suitable answer, e.g.** River flooding; sea flooding; winter storms; snow and ice; or drought (e.g. the 2010–12 drought).
   b) **Any suitable answers, e.g. for the 2010–12 drought:**
   Economic and social impacts: farmers struggled to provide water for livestock and harvest crops; low reservoir levels and hosepipe bans affecting 6 million consumers.
   Environmental impacts: groundwater and river levels very low, affecting aquatic ecosystems; wildfires spread.
   Management strategies: hosepipe bans were introduced; water meters were installed to monitor usage; water companies fixed leaking pipes; water-saving devices were encouraged; education about water use; improved wastewater recycling; new reservoirs and pipelines considered; desalination plants considered.
3. **Any suitable answer, e.g.** Flood warnings are increasing; greater frequency of extreme weather events; some animal and plant species are migrating north; vineyards are now viable in wider areas of the UK (e.g. Wales, Hampshire and West Sussex); jet stream has more variable loops, which are wider/deeper.

Page 21 Quick Test

1. **Any suitable answers, e.g.** Ice cores; marine sediment cores; pollen analysis
2. **Any suitable answers, e.g.**
   Physical causes: orbital changes (Milankovitch cycles); solar output (changes in the Sun's energy); asteroid/comet collisions
   Human causes: use of fossil fuels producing greenhouse gases (e.g. power generation), transportation; agriculture (e.g. methane from cattle, rice paddies); deforestation (e.g. tree removal reduces natural carbon sequestration); methane release from melting permafrost and ocean floors (e.g. due to anthropogenic global warming).
3. a) Scenarios
   b) **Any suitable answer, e.g.** Unpredictable physical factors, e.g. solar radiation, volcanic eruptions that can lead to atmospheric cooling; unpredictable human factors, e.g. too soon to assess impact of target-setting, economic trends, new low-carbon technologies; trends are difficult to spot with a relatively short period of accurate observations; uncertainty about the influence of upper atmosphere, sea currents and deep ocean temperatures.

Page 23 Quick Test

1. a) not adaptation   b) not mitigation   c) not adaptation
   d) not adaptation   e) not adaptation   f) not mitigation
2. Surface waters in the eastern Pacific, near the west coast of Central and South America.

Page 24: Tectonic Hazards 1

1. Convection currents **[1]** in the mantle **[1]**.
2. **Diagram should include at least four of:** convergence, denser oceanic plate, less dense continental plate, melting plate, rising magma, fold mountains. For example:

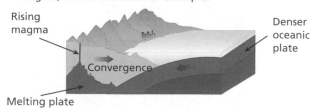

Rising magma

Denser oceanic plate

Convergence

Melting plate

[4]

3. **Diagram should include at least four of:** divergence, rising magma, underwater volcanoes, new rock formed, spreading sea floor, oceanic ridge. For example:

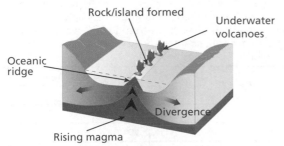

[4]

4. **Diagram should include at least:** direction of movement, friction, faults, crust not destroyed. For example:

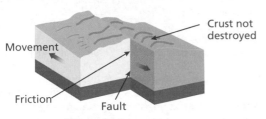

[4]

Page 24: Tectonic Hazards 2
1. LIC – lower income country [1]; HIC – higher income country [1]
2. A – True [1]; B – True [1]; C – True [1]; D – True [1]; E – False [1]
3. **Answer may include references to the factors of:** Degree of preparedness, HIC v LIC, population density, quality of buildings, poverty, depth of focus, etc. [4], and include named examples (such as Tohoku, Japan 2011 – high magnitude and high death toll; Haiti 2010 – lower magnitude and high death toll). [2]

Page 24: Tectonic Hazards 3
1. **Any two from:** Volcanic ash plume at 11 000 metres; fine-grained ash was a hazard to air traffic; lava flows; severe flooding; damage to roads, bridges, water supplies and livestock. [2]
2. **Any suitable answer, e.g.** Continued research into improving forecast methods for eruptions; improved warning and evacuation procedures; the construction of safer homes; people moved away from the most vulnerable areas around a volcano; further research into the effects of ash eruptions on air traffic; the construction of dams to hold back lahars; financial support made available to affected farmers and residents. [Up to 4 marks]

Page 24: Global Circulation System and Atmospheric Hazards
1. B [1]
2. **Any three suitable features, e.g.** The tropical storm has a circular shape; the cloud is spinning in an anti-clockwise direction inwards towards the centre (which means it is in the Northern Hemisphere); there is a vortex; the storm has an eye. [3]
3. a) Cyclone [1]
   b) Hurricane [1]
   c) Typhoon [1]
4. The Coriolis force (Coriolis effect) makes low-pressure systems spin anti-clockwise in the Northern Hemisphere [2] and clockwise in the Southern Hemisphere. [2]

Page 25: Tropical Storms – Case Study
1. Prediction [1]; protection [1]; planning [1]
2. **Any suitable example, e.g.**
   Name: Typhoon Haiyan [1]
   Where: The Philippines, South-East Asia (capital: Manila) [1]
   When: November, 2013 [1]
3.

Impacts	Economic	Social/ Political	Environmental
Homelessness		✓	
Factories and other businesses closed or inaccessible due to damage to transport infrastructure	✓		
Waterborne diseases		✓	
Damage to ecosystems			✓
Schools closed for weeks		✓	

[1 for each correct row; accept 'Environmental' for waterborne diseases]

4. a) P [1]   b) S [1]   c) S [1]
   d) P [1]   e) P [1]   f) S [1]

Page 26: UK Climate and Extreme Weather Events
1. **Any suitable answers, e.g.** 2010–12 drought [1]
2. Higher ground in the west has greater precipitation (relief rainfall) and lower temperatures [1]. Consequently the east is in a rain shadow, receiving much lower precipitation. [1]
3. **Any suitable hydro-meteorological hazards linked to the water cycle, e.g.** River flooding; sea flooding; winter storms; snow and ice; drought [3]
4. **Any suitable answers, e.g. for 2010-12 drought:** Blocking highs (slow-moving anticyclones) led to very dry winters in 2009–10 and 2011–12 [1]. East winds were common, bringing in dry continental air [1]. Significantly lower precipitation than normal [1]. Climate change may have contributed. [1]

Page 26: Climate Change 1
1. D [1]
2. Anthropogenic factors [1]
3. **Any suitable causes, e.g.** Use of fossil fuels producing greenhouse gases (e.g. power generation; transportation); agriculture (e.g. methane from cattle; rice paddies); deforestation (e.g. tree removal reduces natural carbon sequestration); methane release from landfill sites. [3]
4. Glacial – Extremely cold periods of glacier growth
   Interglacial – Warmer periods of glacier retreat
   Quaternary – Last 2.6 million years, including the 'Ice Age'
   Proxy measure – Indirect ways to find out average temperatures from the past
   [3 if all correct; 2 if two correct; 1 if one correct]

Page 27: Climate Change 2
1. B [1]
2. Warmer seawater lacks nutrients [1] so the marine food web collapses, reducing fish stocks [1] and damaging the fishing industry.
3.

	Western Pacific	Eastern Pacific
Surface waters	Surface waters cooler than normal.	Surface waters warmer than normal.
Pressure	High-pressure systems develop where low pressure is the norm.	Low-pressure systems develop where high pressure is the norm.
Rain	Below average rainfall, causing droughts and wildfires.	Well above average rainfall, causing flooding and landslides.

[2 marks if all correct; 1 mark if at least one row correct]

4. a) A [1]   b) M [1]   c) A [1]
   d) M [1]   e) M [1]   f) M [1]

Pages 28–41 Revise Questions

Page 29 Quick Test
1. **Any suitable answers, e.g.**
   Large-scale: tropical rainforest, hot desert, temperate deciduous forest
   Small-scale: pond, hedgerow
2. Food chains show simple relationships between elements within an ecosystem, whereas food webs show complex interrelationships.
3. Biotic refers to living parts of an ecosystem; abiotic refers to non-living parts of it
4. Bird eats seed; bird dies; fungi break down remains; trees reabsorb nutrients from fungi.
5. Carbon dioxide and sulphur dioxide **(also accept carbon monoxide)**

Page 31 Quick Test
1. Along the Equator and between the Tropics of Cancer and Capricorn
2. Mountains, coasts and rivers
3. Western Europe, eastern USA, eastern China, Japan

Page 33 Quick Test
1. High altitude, low temperatures
2. Heathland
3. West Lancashire coastal plain
4. Farming, settlement and industry

Page 35 Quick Test
1. Hot and wet all year round
2. Trees (e.g. mahogany and kapok) have buttress roots; pitcher plants catch insects.
3. **Any suitable answers, e.g.** Mining; settlement; roads; HEP; farming

Page 37 Quick Test
1. Hot and dry all year round
2. Saguaro cactus has no leaves to cut down transpiration and wide roots to catch rainwater; the quiver tree has fleshy leaves to store moisture
3. **Any suitable answers, e.g.** Climate change; population growth; overgrazing; over-cultivation
4. **Any suitable answer, e.g.** Encouraging less water use among local people; developing planning laws that restrict the size of buildings; planting trees to stabilise sand dunes and also to provide shade, fuelwood and fodder for animals; encouraging the use of drip irrigation

Page 39 Quick Test
1. **Any suitable answers, e.g.** Great Barrier Reef; Red Sea; Caribbean Sea
2. They act like giant forests, absorbing carbon dioxide as they grow.
3. They have broad leaves and shed their leaves.
4. **Any suitable answer, e.g.** Timber; recreation; farming

Page 41 Quick Test
1. Cold and dry in winter, mild and dry in summer
2. Soil with a permanently frozen layer
3. Benefits – jobs, profits; drawbacks – environmental destruction
4. Tourists being attracted to new and hostile places

Pages 42–45 Review Questions

Page 42: Tectonic Hazards 1
1. An area that develops above a mantle plume of rising heat from deep in the Earth [1]. Magma generated rises and works its way through a thin section of crust to the surface [1].
2. **Any suitable answers, e.g.** Poor quality housing; poor infrastructure making it harder to reach affected people; less money to protect people (e.g. earthquake-proof buildings); less money for responses (e.g. providing food and water); poor healthcare and facilities. [3]
3. A – Crust [1]; B – Mantle [1]; C – Inner core [1]; D – Outer core [1]
4. **Any suitable answers, e.g.** Building design; monitoring; firebreaks; training of emergency services; education of people in evacuation procedures; planned evacuations; survival kits; overseas aid [6]

Page 42: Tectonic Hazards 2
1. Richter scale [1]
2. **Any suitable answers with references to the differences between the following in HICs and LICs:** Number of deaths; number of homeless; building damage; effect of communications problems on people's lives; quality of construction of housing; quality of infrastructure influences ease of access to affected areas; ability to provide healthcare; population density; need for overseas aid. [6]

Page 42: Tectonic Hazards 3
1. **Diagram should include at least four of:** Crater, gentle slopes, low wide cone, lava layers, runny lava. For example: [4]

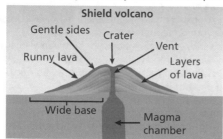

Shield volcano
Gentle sides / Crater / Vent / Runny lava / Layers of lava / Wide base / Magma chamber

2. Ash cloud – Blocks out the Sun, causing suffocation and health problems
   Lahar – Mudslide including rock debris and water
   Pyroclastic flow – Torrent of hot ash, rock, and gases and steam
   Extinct volcano – Volcano nobody expects to erupt ever again
   Dormant volcano – Volcano that has erupted in past 2000 years but is not currently active
   [4 if all correct; 3 if three correct; 2 if two correct; 1 if one correct]

Page 43: Global Circulation System and Atmospheric Hazards
1. Energy (and moisture) [1]
2. The intense low pressure [1] of a tropical storm creates a 'dome' of seawater [1] and a storm surge occurs that causes coastal flooding. [1]
3. Warmer oceans expand, so storm surges may be worse [1]. Climate change may alter the distribution of tropical storms, and their frequency and intensity may increase [1], but the evidence for this is inconclusive [1].
4.

Country	Name used for Tropical Storms
Mexico	Hurricanes
Philippines	**Typhoons**
Australia	Cyclones
Bangladesh	**Cyclones**

[1 for each correct column]

Page 43: Tropical Storms – Case Study
1. D [1]
2. When people choose to stay in their homes [1], even though they have been warned that they may be under threat from a natural hazard [1].
3. Human factors can make matters worse (3 Ps). Prediction – high-level warnings may be delayed (poor quality of governance and communications) [1]; Protection – communications infrastructure may be too vulnerable (e.g. it failed in the Visayas in the Philippines) and disease due to the lack of food, water, shelter and medication [1]; Planning – even though many people may be warned, they may choose to stay in their homes (inertia). [1]
4. Economic impacts are much lower in Bangladesh (poorer country) than in Florida, USA (richer country) [1]. The value of homes and businesses is greater in a richer country [1]. The death toll is much smaller in absolute and relative terms in the USA than in Bangladesh [1]. This is owing to the greater capacity of a richer country to cope with the impacts than a poorer country (richer countries tend to have greater resilience) [1]; 3 Ps more effective in the USA than in Bangladesh. [1] [Up to 4 marks].

Page 44: UK Climate and Extreme Weather Events
1. **Any suitable answers, e.g.** Economic and social impacts: farmers struggled to provide water for livestock and harvest crops; low reservoir levels and hosepipe bans affected 6 million consumers. [1] Environmental impacts: groundwater and river levels were very low, affecting aquatic ecosystems; wildfires. [1]
2. **Any three from:** Hosepipe bans; water meters installed; water companies fix leaking pipes; water-saving devices; education about water use; improved wastewater recycling; new reservoirs and pipelines considered; desalination plants considered (but high cost). [3]
3. The prevailing winds are westerlies from the Atlantic Ocean [1], and the relatively warm, moist air from the North Atlantic Drift or Gulf Stream [1] creates a dominant maritime influence [1].
4. Place A – Mild winters, cool summers
   Place B – Mild winters, warm summers
   Place C – Cold winters, cool summers
   Place D – Cold winters, warm summers
   [3 if all correct; 2 if two correct; 1 if one correct]

Page 44: Climate Change 1
1. E [1]
2. Milankovitch cycles [1]
3. It allows short-wave radiation in (e.g. light) [1], but greenhouse gases prevent the escape into space of some of the heat created (long-wave infrared radiation) [1].
4. **Any two from:** Unpredictable physical factors, e.g. solar radiation / volcanic eruptions which can lead to atmospheric cooling; unpredictable human factors, e.g. too soon to assess impact of target-setting / economic trends / new low-carbon technologies; trends are difficult to spot with a relatively short period of accurate observations; uncertainty about the influence of upper atmosphere / sea currents / deep ocean temperatures. [2]

Page 45: Climate Change 2
1. B [1]
2. Warmer temperatures will lead to higher evaporation from oceans, seas, rivers and lakes [1]. Consequently there is a greater amount of water in the atmosphere for precipitation [1].

3. Seawater absorption of the increased carbon dioxide in the atmosphere [1] will lead to ocean acidification, which is harmful to coral reefs [1]. Their capacity to reproduce declines and coral communities decline in quality and quantity [1].
4. global; harmful; more; acidification; expansion; methane
[3 if all correct; 2 if at least four correct; 1 if at least two correct]

Pages 46–47 Practice Questions

Page 46: Ecosystems and Balance
1. **Any suitable answer, e.g.** Pond; hedgerow [1]
2. A community of biotic (living) [1] and abiotic (non-living) components that create an environment. [1]
3. Food chains show simple relationships between organisms [1]; food webs show more complex relationships. [1]
4. Plants grow and are eaten by herbivores [1]; the herbivores are then eaten by carnivores [1]; the carnivores die and decompose [1]; the decomposed remains are reabsorbed by the trees [1].

Page 46: Ecosystems and Global Atmospheric Circulation
1. High pressure [1]
2. **Any two from:** Mountains; coasts; rivers [2]
3. They are found in a wide belt encircling the Earth, roughly following the Equator and for the most part between the Tropics of Cancer and Capricorn [2]. The Amazon in the northern part of South America and the Congo basin in the centre-west of Africa are the largest rainforests. [2]

Page 46: Ecosystems in the UK
1. **Any suitable answer, e.g.** Wirral peninsula; Breckland (East Anglia) [1]
2. Heather [1]
3. Biodiversity [1]
4. Moorland, heathland, wetland and woodland
[3 if all correct; 2 if two correct; 1 if one correct]

Page 46: Tropical Rainforests
1. Only trees above a certain height are felled [1], thus leaving smaller trees to attain maturity [1].
2. Commercial means farming for profit [1], whereas subsistence means farming to feed oneself [1].
3. The layer above the dark forest floor is the shrub layer of small bushes [1]. Above this is the under canopy, which is a dark, branchless layer of tree trunks [1]. The penultimate layer is the canopy where most of the life can be found [1]. The final top layer is the emergent trees [1].
4. Hunter-gatherers collect edible plants and catch wild animals to eat [1]; soils can be fertilised through small-scale shifting cultivation of crops such as manioc and cassava [1]; trees provide fuelwood and building materials [1]; medicines and hunting poisons can be extracted from a wide variety of plants and animals [1].

Page 47: Hot Deserts
1. Infrastructure [1]
2. Hydroelectric power [1]
3. **Any suitable answer, e.g.** They are found in two broad belts encircling the Earth [1], between 15 and 30° north and south of the Equator [1]. Examples include the Mojave in south-western USA and the Sahara in northern Africa [1].
4. **Any suitable answers, e.g.** Encouraging less water use among local people [1]; developing planning laws that restrict the size of buildings [1]; planting trees to stabilise sand dunes [1]; encouraging the use of drip irrigation in farming [1].

Page 47: Coral Reefs and Deciduous Woodlands
1. Interdependence [1]
2. Great Barrier Reef [1] found off the north-eastern coast of Queensland, Australia [1].
3. When fish die, they decompose at the bottom of the sea (or are eaten by another predator that then excretes their remains onto the sea floor) [1]. This organic material is then re-absorbed by bacteria and algae [1], which return the nitrates back to the system [1].
4. Winter sees the highest rainfall, with places in the USA receiving over 100 mm per month [1]. Daily temperatures are usually between 5 and 7° C in December and January [1] and are usually between 15 and 25° C in June and July [1].

Page 47: Polar and Tundra Environments
1. High latitudes (60–70° north for tundra). [1]
2. Extraction creates huge numbers of jobs [1] but destroys habitats [1].
3. Moderate summer temperatures of 20–25°C [1]. Winter temperatures can plunge below –40°C [1].
4. Plants such as the Arctic poppy [1] survive by developing adaptations such as shallow roots [1] and flowers that track the path of the Sun [1].

Pages 48–65 Revise Questions

Page 49 Quick Test
1. Metamorphic
2. Fast: (any one from) rockfall; avalanche; mudflow; landslide; slump
   Slow: (any one from) solifluction; soil creep
3. Wet

Page 51 Quick Test
1. Moraine
2. Angular / Jagged
3. At the side of a glacier / Along the side of a U-shaped valley
4. Smoothed, polished and scratched

Page 53 Quick Test
1. Between two corries
2. **Any two from:** Truncated spur; hanging valley; waterfall; lateral moraine
3. Abrasion as the glacier flowed over the rock
4. Ice moved more quickly as it passed over the top of the drumlin and streamlined the less-resistant mound of moraine

Page 55 Quick Test
1. Distance of open water over which a wave or wind has travelled
2. Steep (and high)
3. At the foot of a cliff where waves hit it most
4. An eroded area between headlands

Page 57 Quick Test
1. It gets smaller and rounder towards the sea
2. Longshore drift
3. Behind a spit
4. Dragging/Rolling of particles by waves along the seabed

Page 59 Quick Test
1. Rock armour
2. It is reflected
3. Soft
4. Sand/Material is taken offshore by waves

Page 61 Quick Test
1. confluence
2. The difference in minutes/hours between peak rainfall and peak discharge
3. Something that stops or blocks rainfall from reaching the ground
4. The cross-sectional area of the channel, and the water velocity

Page 63 Quick Test
1. Material carried in the river
2. In front of waterfalls
3. Channel sides / River banks
4. Alongside channels on floodplains

Page 65 Quick Test
1. Clearing a river channel
2. Hard
3. They result in more areas of concrete / tarmac roads / paths
4. Wetter and stormier

Pages 66–68 Review Questions

Page 66: Ecosystems and Balance
1. **Any suitable answer, e.g.** Tropical rainforest; hot desert; temperate deciduous rainforest; boreal/taiga forest.
2. An organism that creates its own food [1] through photosynthesis [1].
3. An organism that eats other organisms [1] and can be herbivore or carnivore [1].
4. An organism that breaks down [1] the dead remains of other organisms [1].

Page 66: Ecosystems and Global Atmospheric Circulation
1. The Equator [1]
2. The Tropic of Cancer [1]
3. The Tropic of Capricorn [1]

4. They are found in a wide belt encircling the Earth in what are known as the high latitudes – areas on and a little south of the Arctic Circle [2]. Much of northern Russia, Canada and Iceland have tundra-type environments. [2]

Page 67: Ecosystems in the UK
1. Wetland [1]
2. **Any suitable answer, e.g.** Rushes [1]
3. Moorland [1]
4. Rich soil [1]; flat relief [1]

Page 67: Tropical Rainforests
1. They attract insects using scent glands [1] and then catch them using a slippery flower before digesting them [1].
2. Tourism that uses the beauty of the environment itself as the sole attraction [1] and ensures the protection of the environment [1].
3. **Any suitable answers, e.g.** Tree-labelling schemes; debt reduction; replanting schemes; selective logging [2]
4. **Any suitable answer, e.g.** Schemes such as that at Tucuruí in Brazil [1] created huge amounts of energy for a growing economy [1] but also flooded huge areas of forest [1]. The dam also encouraged further deforestation around it for farming and settlement [1].

Page 68: Hot Deserts
1. Colorado [1]
2. Irrigation [1]; application of organic mulch [1]
3. **Any suitable answer, e.g.** The saguaro cactus [1] has no leaves to cut down moisture loss through transpiration [1]. It has shallow but wide-ranging roots so that it can take advantage of brief rainstorms [1].
4. **Any suitable answer, e.g.** Fennec foxes [1] are nocturnal – they are active at night, which allows them to avoid the extreme temperatures of the day [1]. The fennec also has large ears which act as radiators, allowing it to lose heat [1].

Page 68: Coral Reefs and Deciduous Woodlands
1. Temperate [1]
2. By having broad leaves [1]
3. **Any two from:** Industrial pollution of rivers; global warming; breaking off pieces of coral [2]
4. **Any suitable answer, e.g.** Ecotourism that seeks to use the reef as the attraction itself [1]. Controlling amounts of industrial waste and sewage released into marine ecosystems [1]. Action on global warming through controls on the amount of fossil fuels burned [1]. Setting up marine parks to control tourist numbers [1].

Page 68: Polar and Tundra Environments
1. Permafrost [1]
2. Antarctic Treaty [1]; Madrid Protocol [1]
3. The Arctic fox [1] develops a thick coat to protect against the cold [1]. The Arctic hare has small ears to reduce heat loss [1] and white fur to avoid predators [1].
4. **Any three from:** Oil; bauxite; coal; fish [3]

Pages 69–73 Practice Questions

Page 69: Glaciation 1: Rocks, Weathering and Mass Movement
1. Physical [1]
2. Chalk is a porous [1] rock, which means that water passes into and through it [1]. (If the chalk is saturated or frozen, surface streams can appear.)
3. They are formed through exposure of an igneous, sedimentary or metamorphic rock [1] to great heat, pressure or chemical action [1]. The depth below the surface at which this happens affects the characteristics of the rock [1].
4. **Any suitable answer with four points including comparison, e.g.** North-west Scotland is composed of ancient rocks [1] and was uplifted to form very high mountains [1], whereas south-east England is composed of relatively young rocks [1], which have undergone little disturbance/uplift [1]. South-east England is composed of sedimentary rocks [1], which are susceptible to weathering and erosion [1], especially by running water [1]. North-west Scotland is dominated by metamorphic rocks [1]

with igneous intrusions [1]. These are both impermeable [1] and highly resistant to weathering and erosion [1]. **[Up to 4 marks. Credit could also be given for suitable references to isostatic readjustment]**

Page 69: Glaciation 2: Ice Processes
1. Lowland areas [1]
2. **Any two from:** Material at the base of a glacier can be washed away in meltwater [1] or be worn away [1] so no further abrasion could take place [1]. **[Up to 2 marks]**
3. **A:** En-glacial moraine [1]; **B:** Terminal moraine [1]; **C:** Sub-glacial moraine [1]
4. **Any suitable answer with explanation, e.g.** Harder rock is more resistant to erosion [1] so will wear away less easily [1]. If the bedrock is less hard than the material at the base of a glacier, it will be eroded [1]. If it is harder than the particles at the base of the glacier, it will not be eroded [1] but will wear away the particles themselves [1]. **[Up to 3 marks]**

Page 70: Glaciation 3: Glacial, Physical and Human Landscapes and Post-Glacial Change
1. At the position marking the greatest extent of the ice, i.e. the furthest point reached, such as at the end of a valley [1]
2. It would be flatter/smoother [1], whereas it would have been rocky/irregular/rough when the ice retreated [1].
3. **Any three from:** Lakes for sailing/kayaking/watersports; higher, steeper areas allow for skiing; steep/rugged rock faces allow climbing; beautiful scenery attracts sightseers/painters/writers; hills/mountains attract walkers. [3]
4. **Any suitable answer with comparison, e.g.** On the valley floor soils will be thicker/deeper [1] because it is flat [1] compared with the sides, which can be very steep and so soil is not stable [1]. The soils on the valley floor will be more fertile [1] because streams wash sediments [1] from the valley sides. Valley floor soils will be wetter [1] because water cannot drain away easily [1]. **[Up to 4 marks]**

Page 70: Coasts 1: Weathering and Erosion, and Associated Landforms
1. a) Slide [1]
   b) **Any suitable answer, e.g.** Collapse into a heap; fall apart; be eroded [1]
2. The material that has fallen will remain [1] and protect the cliff foot [1] so the coast will not wear back any further [1].

Page 71: Coasts 2: Transport and Deposition, and Landforms of Deposition
1. Particles approach the shore at an angle or are carried/moved by waves approaching at an angle [1] and are carried/moved up the beach in the swash [1]. They can then be brought back down / brought towards the sea again [1] by the backwash coming straight down [1]. This process is repeated.
2. **Any suitable answer with description [2] and explanation [2], e.g.** A summer beach profile/shape is higher [1] and gentler / less steep [1] than in winter. In summer, the waves are more often constructive [1], adding material [1], but in winter there are more storms [1], with powerful plunging waves [1] that scoop out material / are destructive [1]. **[Up to 4 marks]**

Page 71: Coasts 3: Threats and Management
1. a) **Any one from:** Cheap; dramatic; easily understood [1]
   b) **Any one from:** No means of enforcing behaviour; sign easily damaged/stolen/lost [1]
2. **Any suitable answer which shows the idea of change from… to…, e.g.** Before building, the farmland had a long, even width of beach in front [1] but after building the beach has partially disappeared [1]. Sand/Material has piled up against the updrift side / western side of the jetty [1] but has been eroded / moved away from the jetty / on the downdrift side / to the east of the jetty [1], putting the town at risk [1]. **[Up to 4 marks]**
3. **Any suitable answer, e.g.** The further a wave has to travel after breaking, the less erosive power it has [1]. Beaches absorb wave energy / do not reflect it [1]. Material will be deposited [1], creating even more effective protection.

Page 72: Rivers 1: Drainage Basins

1. **Any two from:** A small catchment area; surrounded by steep slopes; has a lot of tributaries; surrounded by impermeable slopes **[2]**
2. a) 2 hours **[1 – unit must be given]**
   b) 50 mm **[1 – unit must be given]**
   c) 4.5 cumecs **[1 – unit must be given]**
   d) 9 **[1]**

Page 72: Rivers 2: Process and Landscape

1.

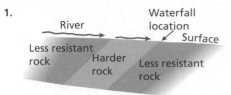

2. **Any suitable answer with four linked points, e.g.** The moving water hits the channel side and loosens material / weakens it **[1]**. Over time, the bank is carved / scooped out **[1]**, leaving the top of the bank overhanging **[1]**. This becomes very weak and collapses **[1]**, making the channel wider **[1]**. **[Up to 4 marks]**
3. Faults are breaks in rocks **[1]** that can expose different bands/ resistance/hardness of rock **[1]**. Softer/Less resistant rocks will be weathered and eroded **[1]**. Over time, rivers will meet a sudden change/step/vertical drop **[1]**.

Page 73: Rivers 3: Management and Flooding

1. a) A dam **[1]**
   b) **General idea:** to control water flow **[1]**
      Explanation – any one from: so that the river never goes over bank/floods; so that there is always enough water for a particular activity, e.g. recreation/farming; so that a lake forms behind for storage. **[1]**
   c) **Any suitable answer with an explanation, e.g.** The force of water over the dam **[1]** causes more erosion below it **[1]**; they are usually made of hard materials **[1]**, which do not fit with the landscape / damage views **[1]**; there is deposition behind the dam from tributary rivers **[1]** so river silt / alluvium is not taken downstream **[1]**; construction and weight of dams **[1]** puts great pressure on the surrounding rocks **[1]**. **[Up to 2 marks]**
2. Chemicals/Nitrates used on fields **[1]** could wash into rivers and damage the water quality **[1]**. **[The same point could be made about animal excrement]**. Ploughing fields loosens soil **[1]**, which can then wash into the river and make the water murky/ less clean **[1]**.

Page 75 Quick Test
1. A city with a population of over 1 million people.
2. 54%
3. Over 10 million

Page 77 Quick Test
1. About 23%
2. **Any suitable answers, e.g.** Favela Bairro Project; upgrading housing; providing pavements; provision of electricity; new sewage systems; legal ownership rights; self-help schemes; low rents; improved transport systems.
3. Local residents are provided with materials like concrete blocks and cement to construct permanent buildings with water and sanitation.

Page 79 Quick Test
1. **Any suitable answers, e.g.** Shops; high-priced apartments; offices; restaurants; yacht marinas; City Airport; watersports; Docklands Light Railway; Thames-side walkways
2. The Crossrail project; HS2
3. The home ground for West Ham United Football Club

Page 81 Quick Test
1. Approximately 85–90%
2. It combines the rings of the Concentric Zone Model with wedges of land use that radiate from the city centre.
3. **Any suitable answers, e.g.** Car sharing; park-and-ride; vehicle restriction zones; integrated public transport systems; alternative fuels; cycleways; walkways

4. A city which produces as much energy as is used (i.e. 'carbon neutral'); a city which supports sustainable lifestyles, such as renewable energy use, local services and jobs, locally-produced food.

Page 83 Quick Test
1. **Any suitable answers, e.g.** London; Merseyside; Greater Manchester; West Midlands; West Yorkshire; Clydeside; Tyneside; Bristol
2. **Any suitable answers, e.g.** Contraception; more women in work; women following careers; improved healthcare reducing infant mortality; financial restrictions on creating large families
3. **Any suitable answers, e.g.** Improved healthcare; new medical treatments and drugs; fewer manual jobs; better diet
4. **Any suitable answers, e.g.** Counter-urbanisation; growth of dormitory villages; increased commuter populations; rural–urban fringe development; second-home ownership; unemployment; declining transport links

Page 85 Quick Test
1. 79%
2. **Any suitable answers, e.g.** Any software companies; any examples of research and development; biotechnology
3. **Any suitable answers, e.g.** Increased car ownership enables people to travel further to shop; growth of out-of-town shopping centres (retail and business parks); decline in rural services; growth of chain stores at the expense of smaller specialist shops; increase in e-tailing; increase in popularity of ethical shopping
4. **Any suitable answers, e.g.** Traffic congestion; land use conflicts; overcrowding at honeypots; erosion to footpaths and trails; litter; seasonal employment

Page 87 Quick Test
1. A leading group of rapidly developing countries consisting of Brazil, Russia, India, China and South Africa.
2. **Any suitable answers, e.g.** Literacy; life expectancy; car ownership; population per doctor; birth rates.
3. **Any four from:** No poverty; no hunger; good health; quality education; gender equality; clean water and sanitation; renewable energy; good jobs and economic growth; innovation and infrastructure; reduced inequalities; sustainable cities and communities; responsible consumption; climate action; life below water; life on land; peace and justice; partnerships for the goals.

Page 89 Quick Test
1. HICs import cheap raw materials from LICs and export more expensive manufactured goods, with LICs doing the opposite. The value of manufactured goods is much higher than primary products, so LICs end up with a trade deficit and need to borrow money from banks in HICs.
2. **Any suitable answers, e.g.** Cheap labour; longer working hours; fewer health and safety laws
3. Long-term aid tries to solve a problem so that it never happens again (e.g. a large dam), whereas short-term aid solves an immediate problem like the lack of housing following an earthquake.

Page 90: Glaciation 1: Rocks, Weathering and Mass Movement
1. Moderate **[1]**
2. Sedimentary **[1]**
3. **Any suitable answer with comparison made, e.g.** All material in a slide moves at the same rate **[1]**, whereas flows move faster at the top and front **[1]**. Slides involve larger particles **[1]**, whereas flows tend to take place with finer particles **[1]**.
4. **Any suitable answer outlining two methods and with linked ideas, e.g.** For freeze-thaw action to take place, water needs to enter cracks/weaknesses in rocks **[1]**, freeze, expand **[1]** and exert pressure to weaken the rock **[1]**. Rainwater is weak carbonic acid **[1]**, which reacts with limestone **[1]** and dissolves it **[1]**. All chemical weathering needs a water supply to maintain the reactions **[1]**. **[Up to 4 marks]**

Page 90: Glaciation 2: Ice Processes
1. Glacier **[1]**
2. **Any suitable answer with two descriptive points or one point explained, e.g.** Polishing **[1]** by finer particles in the ice and scratches/striations **[1]** from bigger/harder particles **[1]**. Smoothing **[1]**. **[Up to 2 marks]**

3. **Any two from:** If melting/thawing was taking place [1] in summer [1]; as glacier retreats [1]; as glacier gets thinner [1].

4. **Any suitable answer, e.g.**

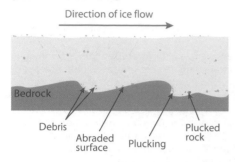

Direction of ice flow

Bedrock

Debris
Abraded surface
Plucking
Plucked rock

[1 for each correct label]

Page 91: Glaciation 3: Glacial, Physical and Human Landscapes and Post-Glacial Change

1. **Any suitable answer with two points, e.g.** Deposits are made of moraine/unconsolidated material/clay with rock fragments [1], which are easily changed/damaged [1], especially by running water [1], so can be eroded or washed away [1]. **[Up to 2 marks]**

2. **Any suitable answer with linked points, stating the place in the corrie and the action, e.g.** Abrasion on the back wall [1] helps to steepen it [1], then to deepen the floor [1]. Ice rotates [1] in the base, smoothing/deepening it [1]. As ice leaves the corrie, abrasion smooths and shapes the threshold [1]. **[Up to 4 marks]**

3. **Any three from:** Weathering and mass movement [1] **[relevant named process gains credit]** could have changed the shape and surfaces **[description related to specific process should be credited]**. Rock faces might be covered by scree/debris [1]. Soil could have formed [1]. Some might be vegetated [1]. **[Up to 3 marks]**

4. **Any four from:** Income from provision of services for tourists [1] **[+1 for example]**; water supply from lakes [1]; hydroelectric power from fast-flowing streams/lakes [1]; forestry on hillsides [1]; arable farming on valley floor [1]; hill sheep farming [1]; mining/quarrying of rocks and minerals [1] **[+1 for example]**. **[Up to 4 marks]**

Page 91: Coasts 1: Weathering and Erosion, and Associated Landforms

1. **Any suitable answer, e.g.** Hydraulic action weakens/loosens the rock of the cliff [1] so that it can fall away [1], giving the cliff an uneven face / causing the cliff to wear back [1]. **[Up to 2 marks]**

2. The material that has fallen will be carried away by waves [1], leaving the cliff vulnerable to erosion / so waves can attack the cliff foot [1]. Therefore the coastline will wear back / the cliff is weakened and may collapse [1].

Page 91: Coasts 2: Transport and Deposition, and Landforms of Deposition

1. **Any suitable answer, e.g.** Dunes are made of loose grains of sand [1] that are dry and so cannot stick together [1]. Sand grains are easily blown / moved by winds [1]. Visitors can disrupt the surface crust, allowing saltation to take place [1]. **[Up to 2 marks]**

2. **Any suitable answer which shows the idea that replenishment is needed as there are always processes at work to remove surface debris from beaches, e.g.** Beaches can be subject to longshore drift that removes material [1] along the shore. Material is taken offshore by destructive waves [1]. Unless more material is added from weathered and eroded cliffs [1] or longshore drift [1], the beach will disappear. **[Up to 3 marks; maximum of 2 marks if only removal process(es) described]**

3. **Any suitable answer which includes the role of the wind and saltation, e.g.** Dry grains of sand are blown by the wind [1] and moved by saltation [1]. They gather at any obstacle [1] (e.g. seaweed) and gradually build up. Tough plants grow [1] (e.g. sea couch or lyme grass) and hold the sand / fix the sand / stabilise the sand [1]. **[Up to 4 marks]**

Page 92: Coasts 3: Threats and Management

1. a) The path is lower than the surrounding dunes [1]
   b) Growth of plants [1]

c) Wooden boardwalk / Pathway constructed [1]

d) **Successful:** It helps to stop people walking on the rest of the dunes [1] so damage will be restricted [1]; People will take less sand away on their feet [1] so dunes will not lose so much sand [1]. **[Up to 2 marks]**
   Unsuccessful: Edges of the path are harder than the dunes [1] so could cause breakages in the dunes [1]; People are focused in one place along the path [1] but they might just as easily walk along the sides of the path and stray on to the dunes [1]; The path provides a natural route for sand to blow through [1] so it will not add to existing dunes / help the dunes to build up [1]. **[Up to 2 marks]**

2. **Any three from:** Sea walls are made of concrete / resistant material [1] to withstand the power of waves and tides [1]. They reflect the energy [1] so that it is returned offshore / so it does not damage the land [1]. They are often curved, which helps to reflect the energy [1]. **[Up to 3 marks]**

Page 92: Rivers 1: Drainage Basins

1. a) Where it begins / Where there is first an identifiable channel [1]
   b) **Any two from:** Glaciers; marshy ground; springs; lake [2]

2. **Any suitable answer with two pairs of linked statements, e.g.** Interception by trees/buildings/vegetation [1] holds water that may then evaporate [1]. Interception by trees/buildings/vegetation [1] holds water that can fall and infiltrate the ground [1]. Rainfall could infiltrate the soil [1] and pass underground / be held as groundwater [1]. **[Up to 4 marks]**

Page 93: Rivers 2: Process and Landscape

1. a) A: waterfall [1]; B: floodplain [1]
   b) Oxbow lake [1]
   c) The source [1]
   d) Headward erosion / Lengthening of stream channel [1]

2. a) At a meander [1]
   b) **Any suitable answer that explains what is happening at each side of the same bend, e.g.** At the outer bend, water flows faster [1] so it has enough energy [1] to erode. Conversely, at the inner bend, water flows slowly [1] so it will deposit some of the load [1].

3. **Any suitable answer which covers the process and explains that the change to the channel occurs at flood or high discharge, e.g.** Water flows sinuously in a channel / moves from one side to the other [1]. Where it hits the bank there will be greater erosion [1]. Eroded material is carried to the opposite bank so that bends are made [1]. The bends get bigger / more exaggerated [1] with reduced land between them. In times of flood / high discharge [1] water may break through, creating a new channel and leaving the meander cut off as an oxbow lake. [1] **[Up to 4 marks]**

Page 94: Rivers 3: Management and Flooding

1. a) **Any one from:** Cheap; uses local material; fits in with the surroundings [1]
   b) **Any one from:** The embankment can get waterlogged and collapse; in the event of over-topping, floodwater cannot easily drain back into the channel. [1]
   c) Vegetation grows on the bank [1], which creates habitats for wildlife [1].
   d) Brick / Concrete [1]
   e) **Any suitable answer with a reason, e.g.** They increase the cross-sectional area of the channel [1] so it can hold more water [1]; They make the banks stronger [1] so they can hold faster-flowing water [1]. **[Up to 2 marks]**

2. **The answer should name the place (region or town/city) and give specific information about it or a particular storm / flood event.**
   For example, for York:
   - It is low lying at 15 m [1] so water from surrounding highlands can gather there [1].
   - It has two rivers running through it, the Foss and Ouse [1], so is doubly vulnerable if discharges rise [1].
   - Upstream from York, the Ouse is joined by the Swale, Ure and Nidd [1] – three rivers from the Yorkshire Dales [1], which is an area of high rainfall [1].

   [Up to 4 marks]

Page 95: Urbanisation

1. a) 10% [1]   b) 54% [1]   c) 60% [1]
2. The uncontrolled expansion of urban areas (towns and cities) [1]. It is considered a problem as it produces deserts of suburban housing [1], reduces productive farmland and destroys wildlife habitats [1].
3. a) **Any three from:** Unemployment; low wages; farming is difficult and unprofitable; few job opportunities; lack of social amenities; isolation; natural disasters. [3]
   b) **Any three from:** More job opportunities; higher wages; better schools and hospitals; better housing and services (like water, electricity and sewerage); better social life; better transport and communications. [3]
4. **Any suitable answer which focuses on four differences, e.g.** The rate of growth (i.e. fastest in LICs); rural to urban migration (reference to push-pull factors); industrialisation (as countries industrialised, people began to migrate to cities for jobs; as this happened in HICs over 100 years ago, they now have large urban populations; LICs are still in the early stages of industrialisation, and so urban populations are still growing); growth of shanty towns (on periphery of LIC cities, but not a feature of HIC cities); primate cities in LICs (LICs often have one primate city rather than a number of large cities of similar size). **[Credit can be given for use of correct geographical terminology, e.g. HIC, LIC, NEE, millionaire cities (growing rapidly in LICs), mega-cities (mainly found in LICs). Credit can be given for relevant examples.] [Up to 4 marks]**

Page 95: Urban Issues and Challenges 1

1. **Any suitable answers, e.g.** Better healthcare; better schools; better housing; jobs; wages; better social services [4]
2. **Any suitable answer, e.g.** Lack of clean water supply; overcrowding; lack of job opportunities; crime rates; lack of services [4]
3. **Any suitable answer, e.g.** Beaches; Sugar Loaf Mountain; statue of Christ the Redeemer; Rio Carnival; sports events [4]
4. **Any suitable answer which refers to two different social challenges [2], with further development of each point needed [2], e.g.** Migration (rapid growth of Rio in recent years from the movement of people into the city from surrounding rural areas putting pressure on services and amenities); housing (many rural migrants begin their life in the city in shanty towns called favelas; these are unplanned and spontaneous, often growing up on poor quality land; favelas are overcrowded, residents have no legal ownership, and houses are built of cheap materials; these are areas lacking clean water and sewage disposal, with no schools or hospitals, and few job opportunities); healthcare (favelas have high levels of disease and illness, with poor quality healthcare – particularly for maternity and care of the elderly); education (nearby schools suffer from poor funding, low enrolment due to poverty, and a lack of trained teaching staff); water supplies (clean water is not available for 12% of the population in Rio's favelas); energy (a shortage of available electricity supply means frequent blackouts for residents); crime (street crime is high, with powerful drugs gangs controlling the favelas).

Page 96: Urban Issues and Challenges 2

1. **Any two from:** By 1980, ships had become too large to sail up the River Thames [1] and the docks became derelict [1]; competition from European ports such as Rotterdam [1].
2. **Any suitable answers, e.g.** Housing; sports facilities; improved transport links; the Queen Elizabeth Park [4]
3. **Any suitable answers, e.g.** Wider use of diesel-electric hybrid buses [1] and the trialling of hydrogen fuel cell buses and combined diesel and biofuel buses [1].
4. **Any suitable answers, e.g.** The Docklands Light Railway (DLR) [1] connects the area to the London Underground system [1]; London City Airport [1] specialises in STOL (short take-off and landing) air travel [1]; high-speed rail link (HS2) from London to Birmingham (and eventually Leeds and Manchester) [1]; Crossrail project from Paddington Station to Reading in the west and Abbey Wood in the east of the city [1]; plans for a new runway at Heathrow [1]. **[Up to 4 marks]**

Page 96: Urban Issues and Challenges 3

1. A greenfield site is one that has never been used for development before [1], whereas a brownfield site has had a previous (now redundant) industrial use [1].
2. **Any suitable answers e.g.** Renewable energy use such as solar panels; insulation; affordable prices and rents; passive energy use; greywater usage for toilets [3]
3. **Any suitable answers e.g.** Shops; banks and offices; chain stores; library; museums; town hall; traffic congestion; overcrowding; high rise buildings; high land values [4]
4. A CBD is located at the centre of the city [1], surrounded by rings of different land uses [1] – the zone of transition is around the CBD [1] then, moving outwards, there are residential areas (lower, middle and upper class) [1]. The rural–urban fringe is on the edge of the city. [1]

Page 96: UK Population and Economic Change 1

1. A diagram that shows the structure of a country's population by age group [1]
2. Birth rates are high [1]; death rates are declining [1]
3. **Any four from:** Pressure on health services; pressure on welfare services; increased demand for retirement homes and nursing homes; economic costs (e.g. state pensions); tax rises for the working population [4]
4. **Any suitable answer with the emphasis on structure and change, e.g.** Fewer children and smaller families [1]; increased working population [1]; greater number of older people [1]

Page 97: UK Population and Economic Change 2

1. **Any suitable answers, e.g.** Bristol; Cambridge [2]
2. **Any suitable answer, e.g.** Growth in car ownership [1]; large range of goods [1]; cheaper prices [1]; less frequent visits to shops [1] **[Up to 3 marks]**
3. **Any suitable answer, e.g.** Shopping choices that give consideration to environmental issues; food provenance; food miles; labour conditions [4]
4. **Any suitable answer, e.g.** Footloose industries are not tied to any one specific location [1]. They are attracted to the M4 corridor because of road links (M4, M5, M25, M11 and M3) [1]; proximity of major airports [1]; labour force from Reading, Slough, Bracknell, Bath and Bristol [1]; ease of access to the Channel Tunnel [1]; nearby universities [1]; access to London [1]; cheap land [1]; attractive countryside [1] etc. Examples include Sony and Hewlett-Packard [1]. **[Up to 6 marks]**

Page 97: Global Development 1

1. **Any suitable answers, e.g.** Birth rate; life expectancy; literacy; infant mortality; access to healthcare [2]
2. **Any suitable answers, e.g.** Gross domestic product (GDP); wages; trade figures [2]
3. **Any suitable answers, e.g.** Deforestation; pollution measures; access to clean water [2]
4. **Any suitable answer, e.g.** Improved literacy provides a route into employment [1]; promotion to a more skilled job [1]; higher wages – to spend on food, education and family [1] and extra taxes for government [1]; greater self-esteem [1]; it can help to promote the use of contraception [1]; it can help to raise the status of women [1] **[Up to 6 marks]**

Page 97: Global Development 2

1. **Any suitable answer, e.g.** The costs involved in purchasing the equipment may be prohibitive [1]; availability of fuel may be restrictive [1]; little availability of replacement parts [1]; ability to provide maintenance may be restrictive [1] **[Up to 2 marks]**
2. **Any suitable answers, e.g.** Faster and greater air travel [1]; high-speed trains [1]; modern containerised shipping [1]; Internet links [1]; satellite phones [1] **[Up to 3 marks]**
3. A TNC is a transnational corporation (or company) [1], e.g. Sony, Nike, Ford, Coca-Cola, Unilever **[Up to 3 marks for examples]**
4. **Any suitable answer, e.g.**
   Positive effects: A TNC can help to invest in infrastructure, such as roads and railways; local companies can become suppliers to the TNC; more job opportunities for the local people; potential to earn higher wages; more opportunities for education and training **[Up to 3 marks]**
   Negative effects: Jobs created may not be secure; the labour force may be exploited and subjected to poor working conditions;

pollution caused by the activities of the TNC; profits may largely be channelled out of the host country; the impact on local people and the economy when a TNC pulls out of a country [Up to 3 marks]

Pages 98–119 Revise Questions

Page 99 Quick Test
1. The UK is unable to grow enough food for its population and there is a great demand for food out of season.
2. Because of a growing population and lifestyles that use more water.
3. It is the driest part of the country but also the area with the highest population and demand.
4. Declining fossil fuel reserves (e.g. gas) mean that the UK has to import more, while more affluent lifestyles mean that households use more energy.

Page 101 Quick Test
1. Geological resources often occur in areas of low demand, while food and water resources are often scarce in countries with harsh climates.
2. Mineral resources such as metal ores are often very valuable, while energy resources allow a country to be energy secure and not rely on expensive imports. Food resources allow a country to have food security.
3. Use of natural resources is rapidly increasing and could lead to potential shortages as their exploitation in accessible locations leads to their depletion. Humans have been forced to exploit resources like oil and gas in pristine wildernesses such as the High Arctic, desert regions and rainforests.

Page 103 Quick Test
1. A country having reliable access to a sufficient quantity of affordable, nutritious food to feed its people.
2. **Any suitable answers, e.g.** Aridity; floods; extreme cold/heat; water availability; soil type
3. **Any suitable answers, e.g.** Civil wars; poor techniques leading to degradation and desertification; poor distribution networks; lack of adequate storage.

Page 105 Quick Test
1. **Any suitable answer, e.g.** The use of new farming methods, such as hydroponics and genetically-modified crops, that supply larger yields; in LICs, examples include high-yielding varieties of cereals and modern farming techniques such as irrigation, synthetic fertilisers and pesticides.
2. **Any suitable answer, e.g.** Eating sources of food that are locally grown and only eating fruit and vegetables that are in season; changing diets to those less reliant on meat and dairy products; using sustainable fish sources and reducing food waste in homes and shops.
3. Ethical consumerism is the purchase and use of sustainable goods, such as food that is locally grown and only eating fruit and vegetables that are in season.

Page 107 Quick Test
1. The ability of a country to ensure access to quantities of safe water to maintain life, social well-being and economic development.
2. In ice caps and glaciers
3. **Any suitable answer, e.g.** Greater demand from agriculture, industry and domestic properties; climate change; over-extraction by man.

Page 109 Quick Test
1. Methods of conservation can be used in the storage of water; expensive schemes, such as water transfer schemes and desalination plants, can also be used.
2. **Any suitable answers, e.g.** Using low-flush toilets; taking short showers; only using washing machines and dishwashers when full.
3. **Any suitable advantages, e.g.** The quality of the water is very good and needs less treatment than river water to make it safe to drink; it stays available during the summer and during droughts when rivers and streams have dried up.
   Any suitable disadvantages, e.g. The use is not sustainable as the groundwater takes a long time to replenish.

Page 111 Quick Test
1. Eastern Europe, including Russia – has large reserves of natural gas and coal; Middle East and North Africa – have large oil reserves.
2. Energy usage is high in countries that have either hot or cold climates due to heating or air-conditioning use; usage is also high in countries with a great annual range of temperature.

Page 113 Quick Test
1. **Any suitable answers, e.g.** The exploitation of oil in harsh climates; shipping the oil out of remote locations; pollution of pristine wildernesses; local and international opposition (e.g. Native Americans, Greenpeace)
2. Wind power is the most widely used renewable in the UK because it experiences windy conditions for much of the year. Nearly 7000 onshore and offshore wind turbines produce almost 10% of the UK's electricity output.

Page 115 Quick Test
1. World reserves of coal, oil and gas are running out and they are also a major source of greenhouse gases that contribute to climate change.
2. Uranium is in plentiful supply and more fuel is produced in a nuclear reaction. Uranium is also much more efficient in producing heat than coal, to turn water into steam to drive turbines. Nuclear power generation produces virtually no greenhouse gases.
3. In HICs, carbon footprints tend to be high because of the lifestyles people lead, using fossil fuels for transport and energy along with diets that largely rely on imported food. Carbon footprints are lower in LICs as lifestyles are more sustainable with less fossil fuel usage and food tends to be locally produced.
4. Energy conservation can be achieved through the intelligent design/building of homes, e.g. the use of insulation in loft spaces and walls, double-glazed windows and larger windows in south-facing walls. Energy-efficient devices like plastic kettles can also help. People can also change habits to conserve energy, e.g. turning off appliances when they are not in use rather than using standby buttons.

Page 117 Quick Test
1. A hypothesis is an idea or explanation for something that has not yet been proved.
2. Primary data is information collected by you and your fellow students. Secondary data is information you have found on a website or in a book (i.e. which someone has collected previously).
3. Quantitative data includes statistics and numbers (i.e. things that can be counted or calculated). Qualitative data is non-numerical and can be subjective (i.e. things such as field sketches, opinions, news articles).

Page 119 Quick Test
1. Synoptic thinking means using the knowledge, understanding and skills that you have learned throughout the course to help answer questions about a new situation that you haven't studied before. Critical thinking is interpreting, analysing and evaluating ideas and arguments based on the information that you are given and your existing knowledge and skills.
2. a) Stakeholders are people or groups with a real interest in an issue or in the development of a decision or policy, e.g. the residents of an area where a new road is to be built.
   b) Key players are people or groups with an interest and a significant degree of control about an issue or in the development of a decision or policy, e.g. the company paid to build the road, or the politician who gives it the go-ahead.
   c) Interest groups are organisations of people with a common cause about which they attempt to influence policies or decisions, without seeking political control, e.g. if the residents of the area where the new road is to be built formed a group, it would be an interest group.
3. 'Top-down' management is where the person/organisation in charge makes all the decisions and tells those lower down exactly what to do. A 'bottom-up' approach takes into account everyone's opinion and the decision is formed by the whole group.
4. Measures that provide people with a degree of control and ownership over problem-solving or decisions.

Pages 120–123 Review Questions

Page 120: Urbanisation
1. **Any simple definition of urbanisation, making reference to an increase in the proportion of people living in urban areas (or towns / cities / built-up areas).** [1]

2. **Any suitable answer with two different reasons. A suitable answer should refer to any 'pull' factor of rural to urban migration, such as industrialisation creating a demand for jobs, better healthcare, better schools, etc. or any 'push' factor of rural to urban migration, e.g. unemployment, difficulties in farming, low wages, etc.** [2]
3. a) False [1] b) True [1] c) False [1] d) True [1] e) False [1]
4. a) Push factor [1]   b) Pull factor [1]   c) Push factor [1]
   d) Push factor [1]   e) Pull factor [1]
5. Greenbelts are bands of land surrounding towns and cities where development is strictly controlled [1]. They are designed to stop urban areas spreading outwards into rural areas [1].

Page 120: Urban Issues and Challenges 1
1. **A suitable answer could include:** Provision of housing for rural migrants (instead of allowing uncontrolled growth of favela housing); upgrading of favelas (by providing pavements, electricity and sewage systems through schemes like the Favela Bairro Project); promotion of self-help building schemes (where local residents are provided with building materials); giving residents legal rights of ownership or low rents on properties (to develop permanent communities); improved transport systems (to give residents better access to work in the city); encouraging creation of businesses like shops and restaurants (to provide employment and allow favelas to be self-contained); improved law and order through pacification programmes (to reduce crime in the favelas); encouraging tourism and creation of tourist-related businesses (to bring additional income opportunities to residents). **[Two different ways that quality of life can be improved need to be identified [2], with further development required [2]]**
2. **A suitable answer could include:** Strength of community; creation of new businesses (e.g. shops, restaurants, tourist-related businesses, cottage industries like pottery); residents sending money home to families in the countryside; recycling (with particular reference to use of building materials and waste recycling); cultural and ethnic diversity; growth in tourist-related activities. **[Three different positive aspects need to be given [3]]**
3. **A suitable answer could include:** Urban sprawl (as the city continues to grow, it encroaches on surrounding rural areas); pollution (air pollution from heavy traffic); traffic congestion (in the city centre); sea pollution (from sewage and industrial waste); waste disposal (particularly in the favelas, many of which are inaccessible to collection vehicles). **[Identification of one environmental challenge [1], with further development required [1]]**
4. **A suitable answer could include:** Favelas being unplanned and spontaneous (often growing up on poor quality land); overcrowding (with many houses crammed into a small area); poor quality housing (often built from cheap materials like wood or corrugated iron); lack of services (like clean water, sewage disposal, electricity supply); poor health (due to lack of clean water and lack of sewage disposal); few job opportunities (leading to poverty); few facilities (such as schools, hospitals and public transport); high levels of crime (particularly street crime, with powerful drugs gangs controlling many areas). **[Reference needed to a range of different living conditions for up to 2 marks, with further development of individual points needed to gain additional marks (up to a maximum of 4 in total).]**

Page 121: Urban Issues and Challenges 2
1. **A suitable answer could include:** Dereliction (as manufacturing industries have declined, particularly in the inner city, much land has been left in a state of dereliction); waste disposal (a large city produces a lot of household and commercial waste for disposal); pollution (atmospheric pollution from industry and vehicles). **[Two different environmental challenges identified [2], with further development of individual points [2]]**
2. **A suitable answer could include:** The Docklands Light Railway (DLR – which connects the area to the London Underground system); London City Airport (a STOL airport providing quick access to the city business areas); improved public transport (new underground station next to the Olympic site). **[Two different transport developments identified [2], with further development of individual points [2]].**

3. **Any suitable answer, e.g.** Historic buildings; monuments; sporting events; cultural events; entertainment **[Credit can be given for named relevant examples]. [Up to 3 marks]**
4. **A suitable answer which includes three social challenges, e.g.** Housing inequalities (provision of more affordable housing); need to improve education standards to correct variation in performance between boroughs; poor access to healthcare for people with poor English language skills; increased in-migration (particularly international migration) attracted by finance and knowledge-based employment opportunities. **[Up to 3 marks]**

Page 121: Urban Issues and Challenges 3
1. a) 17% [1]   b) 77% [1]   c) 87% [1]
2. a) The opportunity of jobs provided by the factories of the Industrial Revolution [1]
   b) **Any two from:** Machinery taking over jobs on farms; surplus labour; agricultural recession following the Napoleonic Wars [2]
3. a) Burgess's model portrays a city in a series of concentric land use zones [1], while Hoyt's model shows the land use zones in a city as sectors or wedges [1].
   b) The zone of transition [1], containing a mix of residential land use and old industrial buildings [1].
4. **A suitable answer could include:** Urban greening (a target of 40% green space of parklands and trees help to meet leisure needs of residents and absorb carbon dioxide from the atmosphere); urban forests (e.g. Adelaide in Australia – they help to meet leisure needs of residents and absorb carbon dioxide from the atmosphere); urban architecture (such as vertical farming, allotments and roof-top gardens, help to increase food production and reduce 'food miles' from imported products); use of brownfield sites for development (to avoid the need to expand into untouched greenfield sites at the edge of the urban area); greenbelt (to maintain an area of green space in easy reach of all residents within the urban area). **[Two different environmental improvements need to be identified [2] with further development of individual points needed [2]]**

Page 122: UK Population and Economic Change 1
1. **Any suitable answer, e.g.** North-east Scotland; Cambrian Mountains; Dartmoor; Exmoor [1]
2. **Any suitable answers, e.g.** More volunteers for charity groups; extended family units [2]
3. Counter-urbanisation describes the movement of people from cities to smaller urban settlements and rural areas [1]. Examples of problems include second-home ownership; creation of dormitory settlements; higher house prices, leading to local people being priced out of the housing market **[Up to 2 marks for problems]**
4. **Any suitable answer with examples of changes such as:** Closure of shops and local services [1]; limited public transport [1]; increasing isolation [1]; fewer newcomers except retirees [1]; more second homes [1]; increasing property prices [1]; more unemployment [1]; lower wages [1]; more jobs of a seasonal nature [1]. **[Up to 6 marks]**

Page 122: UK Population and Economic Change 2
1. A specific location that attracts huge numbers of visitors [1]
2. **Any suitable answer explaining problems such as:** Erosion to footpaths and trails; increase in litter; increase in pollution from cars and coaches **[Up to 4 marks]**
3. **Any suitable answer describing management methods, e.g.** Installation of litter bins [1]; installation of wooden walkways [1]; signed trails [1]; information boards [1]; direction and instruction signs [1]; restricted parking [1]; park-and-ride schemes (e.g. Goyt Valley, Peak District) [1] **[Up to 4 marks]**
4. This trend is known as 'de-industrialisation' [1] and it has been caused by a number of factors including increased mechanisation [1]; competition from overseas [1]; and changes in market demands [1]. **[2 marks can be given for quoting examples from agriculture or manufacturing industry that has declined]**

Page 123: Global Development 1
1. HDI combines life expectancy [1], literacy and income to measure development [1]. It has become popular as it does not rely on one single indicator (usually wealth). [1]
2. **Any suitable answer with two clear reasons to show how an NEE differs from an LIC, e.g.** Country wealth (an LIC is

classified by the World Bank as one with less than $1045 GNI per capita); rate of economic development (NEEs are beginning to experience high rates of economic development, usually with rapid industrialisation); reliance on agriculture (NEEs no longer rely primarily on agriculture in their economy). **[Credit can be given for naming relevant examples.] [Up to 2 marks]**
3. **Any suitable answers, e.g.** Corrupt governments; war and conflicts; colonial exploitation **[2]**
4. **Any three from:** Population increasing rapidly; high birth rates; falling death rates; many young dependants **[3]**

Page 123: Global Development 2
1. **Any suitable answers which include:** A description of intermediate technology (a form of aid, sometimes also known as 'simple' or 'appropriate' technology that does not necessarily use the most up-to-date technology available to improve a situation in an LIC) **[2]**; and for identifying how intermediate technology can reduce the development gap (the difference in standard of living between rich and poor countries) through making improvements to living or working conditions in an LIC **[2]**. **[Credit can be given for reference to examples of intermediate technology aid projects.]**
2. **Any suitable answers, e.g.** Cheap labour; longer working hours; fewer health and safety laws; lax environmental laws **[3]**
3. **Any suitable answer, e.g.** Globalisation refers to the process of countries of the world becoming interconnected and dependent on each other **[1]**. This has come about through international trade, international investment and improvements in communications (e.g. high-tech electronic communication such as teleconferencing) **[2]**.**[Credit can be given for specific named examples]**
4. Tied aid is where one country donates money or resources to another but with conditions attached **[1]**, e.g. the donor country may demand a trade agreement from the recipient country **[1]**.

Page 124: Overview of Resources – UK
1. Greater use of nuclear **[1]** and renewables **[1]** to generate electricity.
2. **Any suitable answer, e.g.** Agriculture is worth £10 billion a year to the UK economy **[1]** and provides the raw materials for the food and drink industry **[1]**, which is the UK's fourth leading industrial sector and worth £108 billion a year **[1]**.
3. **Any suitable answer, e.g.** With increasing affluence, water demand is increasing as new housing is built with more than one bathroom **[1]** and consumers demand labour-saving devices such as dishwashers and washing machines **[1]**. Growth in industrial activity in HICs also results in a greater demand for water **[1]**.
4. Water is transferred from wetter regions to drier regions that have greatest demand **[1]**, using pipelines, aqueducts and rivers **[1]**. Methods of conservation can be used in the home such as using showers, low-flush toilets and 'green' appliances, or water can be recycled using greywater harvesting **[1]**. Opening desalination plants, such as the Beckton one in London, can turn river or seawater into drinking water. **[1]**.

Page 124: Overview of Global Inequalities
1. Europe **[1]**, the USA/Canada **[1]** and Japan **[1]**
2. They are the most developed and industrialised **[1]**. They are also HICs and can afford to buy the resources they don't produce **[1]**.
3. They are generally LICs / are less developed **[1]**, have low levels of industrial development **[1]** and therefore have little use for many natural resources **[1]**.
4. **Any area such as a tropical rainforest, desert or Arctic area may have been studied, e.g.** In tropical rainforests, humans exploit the environment for timber and minerals **[1]**. Large areas of pristine forest have been removed to build mines **[1]**, plant palm oil or soya bean plantations **[1]** and develop cattle ranches **[1]**. This can lead to problems such as deforestation **[1]**, biodiversity loss **[1]**, soil erosion **[1]**, desertification **[1]**, pollution of air/water **[1]** and forced migration of the population **[1]**. Humans have also harnessed the power of water in tropical rainforests to build large multipurpose dams **[1]**. **[Up to 5 marks]**

Page 124: Food 1
1. **Any two from:** Climatic factors (too hot, too cold, too dry, too wet); water availability (for irrigation); soil type/fertility **[2]**
2. **Any two from:** Population size (can the country support its people?); inadequate farming skills; limited level of financial investment a country is able to put into its agriculture **[2]**
3. People in HICs, such as the USA and in Europe, consume the most calories per capita on average **[1]** (over 3000 per day **[1]**) and this is much higher than the recommended daily calorie intake of 2500 for men and 2000 for women **[1]**, leading to a higher risk of obesity **[1]** and other health problems such as cardiovascular disease and diabetes **[1]**. Sub-Saharan Africa has many countries experiencing below average per capita calorie intake with averages of 2200 calories per day for the region, while South Asia has an average of around 2600 **[1]**. **[Up to 4 marks]**
4. **Any suitable answers, e.g.** Climate change is likely to hit the countries of Sub-Saharan Africa the worst **[1]**. Temperatures are likely to rise by more than 2°C **[1]** and this will lead to periods of heatwave and drought **[1]**, causing desertification **[1]** in some areas. Heavy rainfall may cause flooding in others **[1]**. This could result in less land available for food production **[1]**. **[Up to 4 marks]**

Page 125: Food 2
1. The application of water to increase the yields of crops, especially in those areas where water is in short supply. **[1]**
2. The 'Blue Revolution' is the growth of aquaculture **[1]** as a very highly productive way of producing food such as aquatic animals and plants in both salt and fresh water **[1]**. Fish farming, such as shrimp or salmon farming **[1]**, and the gathering of seaweed are all methods of aquaculture **[1]**. **[Up to 2 marks]**
3. **Any suitable answers, e.g.**
Hydroponics is a method of growing plants using nutrient-rich water **[1]**, often using inputs of ultra-violet light **[1]**, such as high-value crops like tomatoes in glasshouses **[1]**. **[Up to 2 marks]**
Aeroponics is a similar way of growing plants in an air or misty environment **[1]** without the use of soil **[1]**. It has been used by NASA as a possible way of feeding astronauts on long space journeys **[1]**. **[Up to 2 marks]**
Biotechnology uses biological processes to develop new products **[1]** that can help feed the hungry by producing higher crop yields **[1]**. It can lower the input of agricultural chemicals into crops **[1]**, resulting in fewer vitamin and nutrient deficiencies **[1]** and reduced allergens and toxins **[1]**. **[Up to 2 marks]**
Appropriate (or intermediate) technology is technology suited to the social and economic conditions of the people using it **[1]**. It is environmentally sound **[1]** and promotes sustainability as it uses locally sourced materials **[1]**. Examples include digging boreholes to supply water to irrigation schemes **[1]** or collecting animal dung to convert into methane for gas stoves to use instead of firewood **[1]**. **[Up to 2 marks]**
4. **Any suitable answers, e.g.** The Gezira Scheme in Sudan **[1]** is an example of a large-scale irrigation scheme where the floodwaters of the Nile **[1]** are used to grow crops like cotton and wheat **[1]**, but large amounts of water are taken out that affect other irrigation schemes downstream **[1]**. The cotton is a valuable export for the Sudanese economy **[1]** but attempts to diversify the type of crops have failed as the country has been affected by civil war **[1]**. **[Up to 5 marks]**

Page 125: Water 1
1. The inability of a country to ensure sufficient access to quantities of safe water **[1]** to maintain life, social well-being and economic development **[1]**.
2. **Any two from:** The supply of water may be limited by low rainfall; arid places may have an absence of groundwater or surface water supplies; rising populations increase the demand for water for drinking, bathing, agriculture and industry. **[2]**
3. **Any suitable answer, e.g.** Climate change is altering patterns of rainfall around the world **[1]**, causing shortages and droughts in some areas **[1]** and leading to desertification **[1]**. In the future, two-thirds of the world's population may face water shortages **[1]**. **[Up to 3 marks]**

4. **Any suitable answer, e.g.** With increasing economic development, the demand for water also rises [1] as domestic use increases through more labour-saving devices like washing machines and dishwashers [1]. Industrial development takes place, leading to a greater demand for water in manufacturing [1] and in electricity generation [1]. Agricultural use of water also increases owing to a greater demand for meat and dairy products [1]. **[Up to 4 marks]**

Page 125: Water 2
1. Schemes that move water from one river basin where it is available, to another basin where water is less available. [1]
2. Removing salt from seawater, river water or groundwater to make it drinkable. [1]
3. **Any suitable answer, e.g.** Groundwater is often found in porous rocks deep underground in aquifers. Groundwater is relied upon heavily in many developing countries, especially in Africa, because it can often be found close to villages and may be over-extracted [1]. Aquifers are often slow to recharge (fill up) [1], so may not always be sustainable, and groundwater supplies can also become contaminated and dangerous to use [1].
4. **Any suitable answer, e.g.** The South–North Water Transfer Project [1] will enable 44.8 billion cubic metres of water per year [1] to move from the Yangtze River in southern China [1] to the Yellow River Basin in arid northern China [1] through a system of canals and aqueducts [1]. The cost of the project is over $60 billion [1] but is likely to cause water shortages in some parts of China and pollution of some rivers [1]. **[Up to 5 marks]**

Page 126: Energy 1
1. The uninterrupted availability of energy sources at an affordable price [1]
2. A lack of access to modern energy services [1]
3. **Any suitable answer, e.g.** A lack of fuel resources available [1] and a reliance on imported fossil fuels [1]
4. North America has vast coal resources but its conventional oil resources are mostly exploited [1]; it is now exploiting non-conventional oil and gas reserves but its huge energy consumption often outweighs supplies [1]. Europe is heavily dependent on energy imports [1] as it has declining fossil fuel supply [1].

Page 126: Energy 2
1. **Any two from:** Via dams that have turbines that are turned by flowing water to create hydroelectric power; from tidal barrages that work in a similar way; via wave-powered generators [2]
2. Solar power has more potential in countries with high annual amounts of sunshine [1]. In the UK, there are not enough sunlight hours in places like Scotland during winter [1] and even with over 650 000 solar power installations, only about 3% of total electricity generation is normally possible [1].
3. **Advantages:** wind is a renewable energy source; there are no fuel costs; no harmful polluting gases are produced **[Up to 2 marks]**
   Disadvantages: wind farms are noisy and may spoil the view for people living near them; the amount of electricity generated depends on the strength of the wind; if there is no wind, there is no electricity **[Up to 2 marks]**
4. **Any suitable answer, e.g.** Agriculture uses oil products to power farm machinery [1], for transport of goods and livestock [1], and in agricultural chemicals such as fertilisers and pesticides [1]. In recent years, agricultural goods like barley, maize and sugar cane have been used to make biofuels [1] that are used as a substitute for oil-based fuels [1]. Rising prices for oil mean a higher price for biofuels and agricultural chemicals [1], making food more expensive [1]. **[Up to 5 marks]**

Page 127: Energy 3
1. Fuels such as coal, oil or gas, formed in the geological past from the remains of living organisms. [1]
2. More normally known as global warming, it is the impact on the climate of the additional heat retained in the atmosphere [1] owing to the increased amounts of carbon dioxide and other greenhouse gases released [1] since the Industrial Revolution.
3. Coal is the dirtiest of fuels (producing 31% of all carbon dioxide) so technology has been developed that removes much of this carbon dioxide before or when the coal is burned [1]. The reason for this is the impact of greenhouse gases like carbon dioxide on climate change [1]. Laws have been passed to cut down the emissions of coal-fired power stations [1].

4. **Any suitable answer with four advantages and four disadvantages, e.g.** [8]

Advantages	Disadvantages
Geographical limitations: nuclear power plants don't require a lot of space. [1] Nuclear power stations do not contribute to carbon emissions or pollution. [1] Nuclear energy is by far the most concentrated form of energy; a lot of energy is produced from a small amount of fuel. [1] Nuclear power is reliable; it does not depend on the weather. [1] We can control the output from a nuclear power station to fit our needs. [1] Nuclear power produces a small volume of waste. [1]	They have to be located near water for cooling. [1] Disposal of nuclear waste is very expensive: it is radioactive so it has to be disposed of in such a way that it will not pollute the environment. [1] Decommissioning (closing down and dismantling) nuclear power stations is expensive and takes a long time. [1] Nuclear accidents can spread 'radiation-producing particles' over a wide area – this radiation harms the cells of the body which can make humans sick or even cause death. [1] They could be targets for terrorism. [1]

Page 127: Fieldwork
For all questions, you need to know the titles and be able to describe the location for both of your enquiries. The answers given below are suggestions for things you may consider depending on the nature of your enquiries.
1. a) and b)
   - Secondary data: cost-benefit data; comparing old and contemporary maps, aerial images, photographs; virtual fieldwork (e.g. Street View; geography.org.uk).
   - Primary data: environmental quality survey (to produce environmental quality index); land use mapping; questionnaires; oral histories; field sketches; photographs.
   a) Specific to physical environment
   Rivers / Coast:
   - Secondary data: Government information such as flood-risk maps from Environment Agency; cost-benefit data; comparing old and contemporary maps, aerial images, photographs.
   - Primary data collection methods for depth, width, wetted perimeter, velocity, gradient, sediment size and shape (roughness) using equipment such as a tape measure, range poles, chain, metre ruler, clinometer, pantometer, hydro-prop with impeller, flow meter, floating object (e.g. table tennis ball, cork, orange, stick), stopwatch, callipers. Physical maps (e.g. geology, drainage, floodplain zones, coastal management zones). As appropriate, profiles or transects might be produced. **[Up to 2 marks]**
   b) Specific to human environment
   Urban / Rural:
   - Secondary data: Government information such as census data from the Office of National Statistics; census data for age profile, unemployment, qualifications, housing tenure, house price or rental survey, Index of Multiple Deprivation (composite indicator), rural deprivation. Human geography maps (e.g. infrastructure maps; Goad maps; neighbourhood zones; planning maps).
   - Primary data collection methods for traffic counts; car age or origin; pedestrian counts; shopping quality (e.g. independents vs. chains); rate of turnover for retail premises; footfall; index of decay; building heights; noise levels or air quality (could use a smartphone app). **[Up to 2 marks]**
2. For both environments:
   - Ease of access physically and legally (e.g. land ownership); distance from school and transport availability; size of area or ward in which representative data can be gathered in time available.
   - Safety: risk assessment and measures to reduce risk; links to hypotheses about the area; ability to make comparisons between contrasting places or to observe changes since earlier investigations took place.

Specific to physical environment
- Rivers / Coast: Manageable size, e.g. small area in which students can walk up and down and, if necessary, be near water or steep slopes or cliffs easily and safely. [Up to 1 mark]

Specific to human environment
- Urban / Rural: Manageable size, e.g. small area(s) of a settlement that can be accessed easily and safely on foot; traffic. [Up to 1 mark]

3. Briefly explain the data collection method. 'Justify' means that you need to explain why it was effective and how it promoted your investigation. State whether the data was quantitative or qualitative; explain the advantages of such data. Describe and explain sampling techniques you used to ensure the reliability of your data, e.g. random, systematic, stratified. If you used GIS, explain why such georeferenced data is a useful technique. State how the data contributed to answering a key question or testing a hypothesis. [Up to 3 marks]

4. Briefly explain the limitations of a data collection method (e.g. why errors may have occurred) and how it might be improved. A good answer should consider questions such as these:
If the sampling technique was a limiting factor, what alternative sampling technique might be used? If the sample size was unrepresentative, could more samples be taken? If the reliability of data is an issue due to when it was gathered, would the findings be more reliable if the investigation was repeated at a different time of day / different day of the week / different time of the year? If the reliability of data is an issue due to inconsistent data gathering (e.g. students not using equipment or techniques in the same way), would more training or practice be helpful? [Up to 3 marks]

Pages 128–131 Review Questions

Page 128: Overview of Resources – UK
1. Any suitable answer, e.g. Organic produce is from a sustainable method of food production [1] that does not use synthetic pesticides or fertilisers [1].
2. Any two from: The UK has large amounts of shale that contains gas and could last for 100 years; it is cheap to exploit compared with gas in the North Sea; gas only produces half as much carbon dioxide compared to coal [2]
3. As the UK relies on fossil fuels and domestic supplies of coal, gas and oil are running out [1], it means it has to import more supplies from other countries [1].
4. Any suitable answer, e.g. There are likely be conflicts between the gas companies and local people in the areas where the gas is drilled [1]. This may be due to the impact of traffic [1], pollution of water supplies [1], the possibility of earth tremors [1] or NIMBY (not in my back yard) attitudes [1]. There may also be conflict between gas companies and environmentalists [1]. [Up to 3 marks]

Page 128: Overview of Global Inequalities
1. Materials such as timber, fresh water or mineral deposits that have an economic value [1]
2. A decline in the number of species in an environment [1] and the number of individuals in each species [1].
3. Any suitable answer, e.g. There is an increasing demand for goods and services from a growing global population [1], especially those in HICs [1]. With increasing development and industrialisation, especially in the NEEs [1], demand for goods and services will increase in those countries, leading to further depletion of resources [1]. [Up to 3 marks]
4. Any suitable answer including an example, e.g. Humans exploit resources for economic gain [1], in order to trade [1], for industrial purposes [1] and to feed the 7 billion people living on the planet [1]. Many pristine areas have been affected by humankind's need for resources, such as tropical rainforests (for timber and metals) [1], oceans (for fish and fossil fuels) [1] and polar regions (for fossil fuels and precious minerals) [1]. [Up to 4 marks]

Page 128: Food 1
1. When there is not enough food for the population of an area, causing illness and/or death through starvation [1]
2. Any suitable answers, e.g. Increase the number of irrigation schemes, including small-scale irrigation projects; restructure farming from subsistence to commercial farming; introduce new

high-yielding varieties of staple crops; introduce better storage and transport systems [3]
3. Any suitable answers, e.g.
Physical factors: soils – fertility influences the type of farming; relief – flat land is preferable for arable farming; climate is the most important physical factor [Up to 2 marks]
Human factors: cost of land; market inertia – farmers may have farmed in a certain way and may be reluctant to make changes; governments (grants, tax barriers and subsidies); technology, e.g. genetically modified crops [Up to 2 marks]
4. Any four from: Farmers are poor and are stuck in the cycle of poverty; lack of investment in agriculture due to poverty; climatic factors including climate change and drought; civil war; price fluctuations leading to unstable markets; food wastage; poor storage and distribution networks [4]

Page 129: Food 2
1. Any two from: Ethical consumerism is the purchase of sustainable goods [1], such as sources of food that are locally grown [1], and only eating fruit and vegetables that are in season [1]. [Up to 2 marks]
2. Any suitable answer, e.g. Reducing food waste and losses in both our homes and in shops could save UK consumers up to £2.4 billion a year. The average UK family throws away enough food for around six meals a week [1]. Buying less food in the supermarket and using goods before they go 'off' would lead to a more sustainable lifestyle [1].
3. Malthus concluded that the rate of population growth was faster than the rate that food supplies could grow [1]. In time, population would outstrip food and some people would starve and the population would decline [1]. Some people may die due to war trying to secure food, while others would die from disease as they were malnourished [1].
Boserup, on the other hand, took a more positive view, writing 'necessity is the mother of invention' [1]. This meant someone will always invent a technological fix for any problem [1].
[Up to 2 marks for each theory]
4. Any suitable answer, e.g. The Organopónicos in Havana, Cuba, is a good example of a local scheme to increase food security among the urban poor [1]. It was introduced after the collapse of the Soviet Union, when Cuba lost its ability to produce enough food for its population [1]. It is an example of urban farming that is a sustainable practice, allowing urban dwellers to produce enough food for their families [1], and can include a variety of activities such as vegetable and fruit growing [1]. Organopónicos take up 3.4% of urban land country-wide, and 8% of land in Havana [1]. They produce over 3 million tonnes of organic food [1] and calorie food intake is back at 2600 calories a day after the threat of hunger was a possibility [1]. [Up to 5 marks]

Page 129: Water 1
1. When there is not enough water to meet all demands as a result of physical conditions [1]
2. A lack of investment in water systems or insufficient human capacity to meet the demand for water [1] in areas where the population cannot afford to use an adequate source of water [1].
3. Any suitable answer, e.g. LICs can improve water security by reducing wastage and evaporation [1], transferring water from regions of surplus to regions of deficit [1] and encouraging small-scale storage schemes at a local level [1]. [Up to 2 marks]
4. Any suitable answer, e.g. Poor sanitation [1], meaning people are exposed to cholera, typhoid and other waterborne diseases, such as diarrhoea [1]. Many rivers, lakes and aquifers are drying up [1] or becoming too polluted to use [1].

Page 129: Water 2
1. Diverting supplies from one river system to another [1]; the building of dams and reservoirs increases the supply of water [1]; a number of countries have opened desalination plants [1].
2. Any suitable answer, e.g. The use of showers and low-flush toilets [1], along with 'green' appliances that use little water [1]. Greywater harvesting [1] is a way of conserving water involving the recycling of water from baths and showers [1], as well as from rainwater from roofs, to use for flushing toilets and other non-drinking purposes [1]. [Up to 3 marks]

3. **Any suitable answer, e.g.** Groundwater makes up nearly 30% of all the world's fresh water [1] and is often exploited in places where there is not enough water to drink [1]. Groundwater quality is usually very good [1] and needs less treatment than river water to make it safe to drink [1], as the rocks through which the groundwater flows help to remove pollution [1]. Groundwater also responds slowly to changes in rainfall [1] and so it stays available during the summer and during droughts, when rivers and streams have dried up [1]. Expensive reservoirs are not needed to store groundwater, and little technology is needed to access it, as it is often drawn from wells [1]. **[Up to 5 marks]**

4. **Any suitable answer, e.g.** The Agra clean water project, India. Agra is a city of 1.3 million people [1]. The city's water supply is dependent on the Yamuna River, which provides a limited supply of polluted, undrinkable water [1]. Those who can afford it purchase bottled water or household filters [1], while people living in one of the city's 432 slums generally either tap groundwater supplies or depend on private tankers, which bring in water from outside of the city [1]. The project covers two areas of slums in Agra [1]. They are not connected to the water network and the groundwater is highly polluted by local industry [1]. The Agra clean water project will combine water testing and an education programme [1] to revive traditional knowledge and systems of rainwater harvesting and water conservation [1] to ensure that around 2500 people have their water supplies improved [1]. **[Up to 5 marks]**

Page 130: Energy 1

1. The amount of energy used [1]
2. a) Between 1970 and 2014, the use of oil dropped from 48% to 0% [1], while the use of gas increased from 8% to 30% [1] and coal dropped from 38% to 30% [1].
   b) 19.3% [1]
   c) Improvements in technology have made the harnessing of these sources possible [1]; investment and subsidies have been made available for the installation of wind turbines and solar panels [1]; less reliance on fossil fuels needed to increase energy security [1].

Page 130: Energy 2

1. Oil is used in agricultural chemicals, such as fertilisers and pesticides [1]. Oil is also used as a raw material in many industries, such as plastics, packaging and textiles [1].
2. **Any suitable answer, e.g.** The manufacturing process has to use a lot of energy and the carbon dioxide emissions from burning the biofuels may be greater than the uptake as the plants grow [1]. Also, there is an argument that the land would be better used growing crops for food, to help relieve shortages in LICs [1].
3. **Any suitable answer, e.g.** LICs may not be able to afford to build hydroelectric plants [1]. If they receive low amounts of rainfall, this may cause water shortages [1]. If there is only flat land, valleys cannot be made into dams successfully [1]. Dams cause environmental damage to rivers [1]. **[Up to 3 marks]**
4. **Any suitable answer, e.g.** The Muppandal wind farm in Tamil Nadu in India [1] has over 3000 wind turbines and is one of the largest in the world [1]. It produces enough electricity to power a million homes [1]. The turbines were purchased by wealthy people who bought them as an alternative to paying high taxes [1]. As India is a heavy user of coal, installations such as Muppandal reduce the amount of carbon dioxide and pollution produced [1].

Page 131: Energy 3

1. A carbon footprint measures the total greenhouse gas emissions caused directly and indirectly by an organisation/person/event/product [1] and is measured in tonnes of carbon dioxide [1].
2. Carbon footprints in HICs tend to be high because of the lifestyles people lead, with a reliance on fossil fuels [1] and a diet that relies on imported food [1]. In LICs, carbon footprints tend to be low as lifestyles are more sustainable [1], less fossil fuel is used [1] and food tends to be locally produced [1]. **[Up to 4 marks]**
3. **Any suitable answer, e.g.** Energy can be conserved in buildings with methods such as the use of insulation in loft spaces [1] and walls [1], double-glazed windows [1] and larger windows in south-facing walls [1]. **[Up to 3 marks]**

4. **Any suitable reasons, e.g.** Public transport and bicycles help to ease traffic congestion; they help to reduce air pollution and carbon dioxide emissions; it is more fuel efficient to carry large numbers of people on public transport than in large numbers of individual cars. [3]

Page 131: Fieldwork

For all questions, you will need to know the titles and be able to describe the location for both of your enquiries. The answers given below are suggestions for things you may consider depending on the nature of your enquiries.

1.
* For quantitative data: scattergraphs, line charts, pie charts, bar charts, histograms with equal class intervals, particular types of bar chart such as population pyramids, divided and cumulative bar and line charts, pictograms, proportional symbols, choropleth maps, isolines, heat maps, dot maps, desire lines and flow-lines. You could also include tools for statistical analysis such as lines of best fit; measures that show the strength of correlation, such as R coefficients; Spearman's Rank correlation; measures of central tendency: median, mean, mode and modal class; measures of spread and cumulative frequency: range, quartiles and inter-quartile range, dispersion graphs; percentages: percentage increase or decrease, percentiles; relationships in bivariate data: trend lines through scatter plots, lines of best fit, positive and negative correlation, strength of correlation; predictions and trends: interpolation, extrapolation; limitations and weaknesses in selective statistical presentation of data; geo-spatial (geolocated or georeferenced) data presented in a geographical information system (GIS) framework; GIS can also be used to analyse spatial data.
* For qualitative data: annotated images, diagrams or field sketches; overlays using GIS or tracing paper; tables comparing quotes or opinions.
[Up to 3 marks]
2. For the technique used, briefly explain what the presentation technique is and how it shows or enhances the information provided by the data and any limitations of it. **[Up to 4 marks]**
3. Structure your answer, considering the following:
Potential risks and likelihood: slipping (high), clothing becoming wet, footwear leaking (high), hypothermia (medium – depends on weather), drowning (water not deep so low risk), allergies, traffic / wildlife (low), local conditions, e.g. stability of slopes. Who may be affected: students, teachers and other accompanying adults, members of the public. **[Up to 4 marks]**
4. A good answer should consider the following:
What are the potential risks? Who may be affected by the risks? What is the level or likelihood of the risk (high, medium, low)? What can be done to reduce/manage the risk? If these measures are taken, how likely is the risk to happen then?
Here is an example of a student's notes for writing a risk assessment:
Potential risk: water in river, nearby streams, traffic near coach;
People affected by the risks: students/staff falling in the water, becoming dangerously cold, slipping or falling on rocks;
Level/likelihood: medium for most, traffic – low;
Risk reduction/management: appropriate clothing and footwear, discussion and advice about risks beforehand, reminders at key times;
Impact of measures: reduced likelihood; if something happens, everyone is prepared and knows what to do. **[Up to 4 marks]**

Pages 132–139 Mixed Questions

1. **Any suitable answer, e.g.** The USA and France are HICs [1] and use large amounts of water for agriculture, industry and domestic purposes [1]. China and India are NEEs [1] and their water usage is increasing due to industrialisation [1] and increasing domestic demand [1]. Mali and Egypt are both arid countries [1] but Mali is an LIC whereas Egypt is an NEE [1] and uses lots of water for irrigation purposes, having plentiful supplies (River Nile) [1]. **[Up to 5 marks]**

2. In general, agriculture mostly occurs rural areas, the main concentrations of arable farming (crops) being in the south and east of the UK [1] and pastoral farming (livestock) in the north and the west [1].

3. Intermediate technology is an appropriate form of technology for the context in which it is operating [1]. It is an improvement on existing simple technology [1] but does not necessarily make use of the most advanced technology available [1]. **[An additional 2 marks are available for description of a named example, such as clay pot 'refrigerators', animal-drawn metal ploughs, etc.]**

4. A settlement where high numbers of commuters reside [1], and with an absence of local services [1], created as a result of counter-urbanisation [1]. **[Credit can be given for specific named examples]**

5. The UK climate can be influenced by relatively dry continental air from land masses to the south and east [1]. In winter, such air can be extremely cold and dry, but in summer, very hot and dry [1].

6. **Any suitable answer with four points or linked ideas, e.g.** Almost all mass movement is made more likely after heavy rain [1]. Water increases the weight [1] of surface material. It makes surface material more mobile / easily moved [1]. Water lubricates the bedrock [1] so that surface material can move more easily across it [1]. **[Up to 4 marks]**

7. **Any suitable answer, e.g.** Too many beetles born in one year means more food [1] for the birds that prey on them [1].

8. Taiga [1]

9. It would have a narrow channel [1], meandering [1] across the floor of the valley.

10. **Causes:** Intense low pressure in a tropical storm creates a dome of seawater, especially around the eye of the storm. As this moves across land, floods occur. High winds cause a 'piling up' of seawater, which makes coastal flooding worse. [2]
Potential impacts: Coastal floods; infrastructure damaged; drownings; if sewage becomes mixed with floodwater, disease can spread. [2]

11. **Any suitable impacts occurring later on, e.g.** Loss of jobs; polluted water supplies; crime (looting) [2]

12. **Any suitable answer, e.g.** Installing solar panels [1] and using energy-efficient devices like LED light bulbs [1] and plastic kettles [1]. Turning off appliances when not in use (rather than using standby buttons) [1] also helps to conserve energy in the home.

13. **Any suitable answers, e.g.** Lack of affordable housing; loss of communities; lack of suitable jobs; disruption during construction phase; communities divided by the London Docklands Development Corporation boundary; people had little say in the process [3]

14. The climate has no extremes [1] and is characterised by relatively moderate temperatures and precipitation [1].

15. **Any suitable answers, e.g.** Traffic restrictions; lower rents; affordable housing; improved security measures to lower crime rates [3]

16. **Any one from:** By shedding their leaves; thick bark [1]

17. a) Sea wall [1] and rock armour/rip-rap [1]
b) **Any suitable answer with linked points for each form of protection, e.g.** Sea walls reflect rather than absorb wave energy [1] so can cause seabed erosion offshore [1]. Sea walls can look ugly / not fit in with surroundings [1] and so put off visitors [1]. Sea walls are expensive to build and/or repair [1], which local councils/authorities may not be able to afford or maintain [1].
Rock armour is not made from local rock [1] so does not fit with the environment / looks ugly [1]. Rock armour can be moved by strong waves / in heavy storms [1] so will need to be replaced/repaired periodically [1].
[Up to 4 marks]

18. The area on the very edge of the city [1] where there are both rural and urban land uses [1] such as agriculture, horse pastures, retail parks and industrial development [1]

19. **Any suitable answer, e.g.** Demand for fossil fuels, especially oil and gas, can mean that producers have to exploit areas that are challenging [1] such as deep oceans [1], polar regions [1] and other remote, difficult and environmentally sensitive areas. The development of technology has enabled this to happen [1]. (Marks are also available for examples such as, in the 1960s, British companies searched for oil and gas in rocks beneath the North Sea [1] and, in Alaska, exploitation of oil resources has taken place in environmentally sensitive areas close to the Arctic Ocean [1].) **[Up to 5 marks]**

20. Approximately 6% [1]

21. Soil erosion occurs [1]

22. Because it has extremely low precipitation [1]

23. It 'fixes' countries into one of two classifications – either developed or not [1] – and does not show how all countries are developing, albeit at different rates [1].

24. **Any two suitable techniques each supported by a diagram, e.g.** Steel reinforcement; floating foundations; shock absorbers; shear walls; fire-resistant materials; evacuation areas; electric shutters **[Up to 2 marks for written description and up to 2 marks for diagrams]**

25. D [1]

26. **Any suitable answer, e.g.** Hoover Dam in the Mojave Desert; Aswan Dam, Egypt [1]

27. **Any suitable answers, e.g.** Amazon; eBay; most large retailers and service providers, e.g. John Lewis [3]

28. **Any suitable answers, e.g.** Congested local roads owing to the high number of visitors at certain times of the year; damage to hedgerows, walls and fences; erosion to footpaths; gates left open; disturbance to livestock; litter [4]

29. **Any suitable answer, e.g.**
Hardraw Force, near Hawes, foot of Buttertubs Pass, North Yorkshire
Hardraw Force has the highest single water drop in England at 100 feet [1]. It falls from a rocky ledge [1] of sandstone with carboniferous limestone above [1] and has a deep plunge pool [1], which is on shale [1].
[Up to 1 mark for accurate location information and up to 3 marks for a coherent and specific description.]

30. a) Seismograph [1]
b) Seismogram [1]

31. **Any suitable answer, e.g.** The Green Revolution is an initiative that has increased agricultural production, particularly in LICs since the late 1960s. The Green Revolution saved around a billion people from starvation [1] and involved the development of high-yielding varieties of cereals [1] (rice and wheat) and new farming techniques [1] including the use of irrigation [1], synthetic fertilisers [1] and pesticides [1]. **[Up to 4 marks]**

32. **Any suitable answer, e.g.** They have large reserves of oil and gas [1], e.g. Saudi Arabia has the world's largest reserve of oil [1], large annual amounts of sunshine for solar power [1] and the windy deserts can be exploited for wind power generation [1]. High oil revenues enable the purchase of technology for renewable energy generation [1]. **[Up to 4 marks]**

33. **Any suitable answer – key elements are:** line of most efficient flow; erosion on outer bend(s); deposition on inner bend(s); pushing out of a bend into a definite meander.

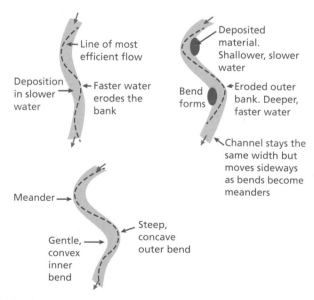

[1 for diagrams; 3 for annotations / labels]

34. **Answers will vary but should consider these points:**
    Reliability of conclusions: For the topic of your enquiry, you need to make judgements about how close your conclusions are to actual changes happening.
    Comments on reliability may be affected by limitations of equipment used; equipment operator errors; choice of data collected; methods of data collection; design methodology / sampling methodology, e.g. number and suitability of sample sites (spatial) and time of day or year (temporal).
    Use these ideas to reach an overall judgement about the reliability of your conclusions. Also evaluate the extent to which your findings would be repeated if the investigation was undertaken at different places (spatial) or times (temporal).
    [9 + 3 SPaG]

dense   pattern or distribution is dense if it shows a high concentration in one area, e.g. dense drainage pattern; dense population **4**

deposition   material being left behind in response to a loss of energy **51, 52–53, 56, 59, 62–63, 64**

depression   another term for low pressure, rising air **14**

desalination   removing minerals from saltwater to render the water suitable for human use **108**

desertification   the process by which previously fertile land becomes desert **37, 101, 103**

destructive margin   a plate boundary that occurs when tectonic plates move together **8–9, 10, 12–13**

destructive wave   high-energy waves which remove material from a beach; backwash is more effective than swash **54, 57**

discharge   the amount of water in a channel passing one place at one time **53, 61, 65**

dispersed   a type of settlement pattern: separate and scattered **4**

distal   furthest from the point of origin; of a spit: the sea end **57**

dormant volcano   an active volcano that is not currently erupting **12**

downcutting   vertical erosion of a channel, making it deeper **62**

downstream   in the direction that a river is flowing **52, 63, 65**

drainage   the removal of water from an area   **4, 6, 53, 60–61, 64**

drainage basin   water from an area that feeds into a river; also known as a river basin or a catchment area **60–61, 64**

dredging   scraping material out of a channel to make it bigger **64–65**

drought   a period of below average precipitation **14, 18–19, 23, 64, 103, 107, 108–109**

drumlin   elongated glacial deposit with a blunt upstream side (stoss end) and tapered downstream side **51, 52**

easting line   one of the vertical lines crossing an OS map from top to bottom; they are called eastings as the numbers increase in an easterly direction **4**

economic activity   a means of making money **58**

economic development   improvement in living standards by creation of jobs **35, 77, 86, 107, 110**

economic water scarcity   the population does not have the monetary means to access available water **106**

ecosystem   a community of interdependent living and non-living elements that create a particular environment   **4, 19, 22, 28–29, 30–31, 32–33, 34, 36, 38–39, 40–41, 59, 64**

ecotourism   responsible travel to areas that sustains the well-being of the local population and environment **35, 38**

elevation profile   shows how far above sea level certain features are **5**

El Niño   Spanish for the 'Christ Child', so-called because around Christmas time every few years surface waters in the eastern Pacific, near the west coast of Central and South America, become warmer than normal. This causes the reversal of wind patterns across the Pacific, causing drought in Australasia, and unseasonal heavy rain in South America. **23**

embankment   artificial bank built alongside a channel; can be made of natural materials or concrete **65**

energy conservation   reduction in the amount of energy used by individuals or groups **114–115**

energy consumption   amount of energy used by individuals or groups **110**

energy insecurity   without access to a secure and affordable energy supply **110**

energy poverty   unable to heat or provide other energy services to homes **110**

energy security   uninterrupted availability of energy sources at an affordable price **110**

en-glacial (moraine)   debris inside the ice **51**

ENSO   El Niño Southern Oscillation; the tendency for oceanic conditions in the Pacific region to swing like a pendulum from El Niño to La Niña and back again **23**

entrainment   the process of material being taken into an agent of transport and erosion, such as a glacier or river **62**

epicentre   the point on the Earth directly above the focus of an earthquake **10–11**

erosion   wearing away of the landscape due to the movement of ice, water or wind **35, 50, 52–53, 54–55, 57, 58–59, 62–63, 64, 85, 101, 103**

erratic   a rock or boulder that differs from the surrounding rock **51**

estuary   tidal river mouth where saltwater from the sea meets freshwater **33, 57**

ethical consumerism   buying only products or services that are produced in a way that does not harm the environment **105**

extinct volcano   a volcano that has not shown an eruption for at least 10 000 years, and is not expected to erupt again in the future **12**

extrapolation   predicting future outcomes based on known facts **7**

extreme temperatures   excessive heat or cold **36, 40**

extreme tourism   travel to remote or unsettled areas which can be dangerous **41**

eye of the storm   if a tropical storm becomes 'cyclonic', it spins so fast that the air around the centre forms a vortex which has an eye 30–65 km wide **15**

fall   mass movement; pieces of rock detach from the face or cliff and travel under gravity to the base; only on resistant rocks **49, 54**

famine   extreme scarcity of food **75, 103, 105**

faulting   breaks in rocks; can be as small as a hairline or huge **63**

favela   an area of makeshift housing in or near a city in Brazil **76–77**

fetch   the distance over which a wave or wind has travelled **54**

flashy   a river that responds quickly to rainstorms **61**

floodplain   low-lying area made of deposited material to the sides of a river channel; flood waters contribute to the formation **33, 60, 63, 64–65**

flood management   measures which are taken to alleviate the negative impacts of flooding on people and their property **6**

flow (mass movement)   when the upper layers of a surface become saturated, often after heavy rain or snowmelt, and move downhill under the influence of gravity **49**

food chain   simple set of connections showing how a small group of organisms are linked; who eats what **28**

food insecurity   being without access to enough food **102–103, 104–105**

food miles   the journey and fuel used from food producer to consumer **85, 98, 105**

food security   having reliable access to enough food **102–103**

food web   a diagram showing how larger groups of organisms are interlinked **28**

footloose industry   industry not specifically tied to one location **84**

fossil fuel   sources of energy formed from the remains of organisms buried millions of years ago; coal, oil and gas **29, 38, 99, 100–101, 110–111**

fracking   drilling the earth and injecting high pressure liquid to extract gas **99, 110**

frequency   the regularity of an event, such as a tropical storm **7, 15, 19, 20, 23**

front   a boundary separating two masses of air of different densities **14, 18**

frost action   trapped water freezes and expands in cracks, joints etc., exerting enough power over time to weaken and break rocks **54**

frost-shattering   the freeze-thaw weathering process **48, 50, 52**

gabions   bundles of rock in a wire mesh cage; used to protect slopes, especially at the coast, on river banks, and roads **58**

geographical enquiry   investigation linking your study in school/college with fieldwork carried out away from school/college **116–117**

geology   the science of Earth's origins, including rocks and minerals **60, 111**

geothermal energy   using heat from deep within the ground (such as near volcanoes) to obtain energy   **9, 113**

GIS   systems designed to capture, store, manipulate, analyse, manage, and present all types of spatial or geographical data; GIS data is geolocated or georeferenced using widely recognised locational methods such as latitude and longitude **5, 7, 117**

glacials   cold periods in ice ages when glaciers and ice sheets grow **20**

glacier   a slow-moving stream of ice formed from accumulation of snow **12, 20, 22, 50–51, 56, 60**

global circulation system   the air in the atmosphere moves around, transferring energy and moisture all around the planet **14**

global warming   the rise in the average temperature of the Earth's atmosphere **21, 38–39**

global weirding   unusual or extreme patterns or trends in weather or climate **22**

globalisation   process whereby the world is becoming increasingly interconnected thanks to mass communications **88**

gorge   steep-sided section of a valley; usually rocky **63**

greenhouse effect   the natural warming of the atmosphere as heat given off from the Earth is absorbed by greenhouse gases such as carbon dioxide; an **enhanced greenhouse effect** is the exaggerated warming of the Earth's atmosphere caused by the emission of greenhouse gases from human activities, resulting in global warming and climate change **20, 22, 114**

greenhouse gas   gas in the atmosphere which absorbs and emits heat (infrared) radiation, such as carbon dioxide, methane and water vapour 20–21, 22, 114

greywater   wastewater from the household that can be recycled 81, 109

groundwater   water held underground in the soil or rock 19, 60–61, 64, 99, 106–107, 108–109

groundwater flow   slow, underground, horizontal movement of water towards the channel 60–61

groynes   man-made structures at right angles to the coast to stop material travelling along the beach by the action of longshore drift 58–59

Gulf Stream   a powerful warm ocean current 18

hard engineering   a range of measures used to manage rivers and coasts which use artificial structures made from materials such as concrete, stone or steel, e.g. wing-dykes, dams, weirs; often more expensive than soft engineering and considered to be less economically, socially and environmentally sustainable 6, 58, 65

headland   an area of often more resistant rock that projects out into the sea 55

heatwave   a period of unusually hot weather 14, 22

helicoidal   like a flattened curve; water takes this motion at meander outer bends, hitting the bank high then moving down and outwards towards the next inner bend 63

HIC   higher income country 10–11, 12, 17, 74, 86–87, 88, 100, 102, 114

high pressure   weather system of descending air, creating stable conditions (lower winds; few clouds; little rain) 14, 23

histogram   data grouped into ranges and plotted as bars 7

honeypot   a location attracting a large number of tourists who, owing to their numbers, place pressure on the environment and local people 85

hot spot   an area of unusually high heat flow in the Earth's mantle, often resulting in volcanic activity (e.g. Hawaii)   9, 12

Human Development Index (HDI)   a measure of development taking account of several indicators 86

hunter-gatherers   tribespeople who live by hunting and harvesting wild food 35

hybrid vehicle   one which uses two or more types of power, such as internal combustion engine plus electric motor 115

hydraulic action   an erosive process which involves the pressure of water in rivers or at the coast hitting a surface, compressing air in any cracks or faults, resulting in the wearing away and removal of rock over time 55, 62–63

hydroelectric power (HEP)   producing electricity using water power 35, 37, 64, 100, 113

hydrograph   a diagram showing the amount of water in a particular river over time; can be produced for minutes, hours, days, months, years 61

hydro-meteorological hazards   weather hazards linked to the water cycle, such as floods   18

hypothesis   a statement or theory that can be tested 116–117

ice sheet   a type of glacier covering large areas 50–51

igneous   rock type; formed from molten magma which cools below (intrusive) or above (extrusive) the surface 48

impacts   the economic, social and environmental effects of an action or decision 6, 8, 16–17, 18–19, 22–23, 59, 64, 99, 101, 103, 106–107, 111, 112, 119

impermeable   a surface that does not allow water to pass through 6, 48–49, 60–61, 64–65

inertia   when people and/or organisations do not take action when faced with the risk of a hazard 16

infrastructure   facilities and supply lines which make modern life possible, such as roads, railways, air and sea ports, water pipes, sewage pipes, electricity cables, telecommunications links, computer networks and Internet access 8–9, 16–17, 37, 41, 84, 87, 88–89, 106

intensity   severity or strength of an event such as a tropical storm 15, 20, 23

intercept   get in the way of; anything that stops rainwater reaching the ground, permanently or temporarily 61, 65

interdependent/interdependence   when two or more organisms rely on one another for their survival 28, 34, 38, 40

interest group   organisation of people with a common cause about which they try to influence policies or decisions, without seeking political control 118–119

interglacials   warmer periods in ice ages 20

interlocking spurs   fingers of land around which rivers flow; they obstruct the view up or down the valley 52, 62

intermediate technology   technology that is appropriate for the developing needs of the community 89

interpolation   predicting values from a limited number of sample data points 7

interrelationship   two-way links between different elements in an ecosystem 28

IPCC   the Intergovernmental Panel on Climate Change 20–21, 22

irrigation   artificial addition of water to land to help crops grow 37, 64, 104, 106, 108

isobars   lines on a weather map joining areas of the same air pressure 6

isohyets   lines on a weather map joining areas of the same rainfall 6

isolines   lines drawn to link different places that share a common value 5, 6–7

jet stream   long, thin current of rapidly moving air 19

key players   people or groups with an interest and a significant degree of control about an issue or in the development of a decision or policy 118–119

lahar a mud flow consisting of volcanic ash and water that runs down the slopes of a volcano, sometimes burying settlements in its path 12–13

landward   facing away from the sea 59

La Niña   Spanish for the 'The Girl'; it occurs when conditions are the opposite of El Niño (colder waters in the eastern Pacific; warmer in the western Pacific) and it therefore causes flooding in Australasia and drought in South America 23

lateral (moraine)   material carried at the side of a glacier and deposited along the side of a valley 51, 52

latitude   latitude lines provide a measure of how far places are north or south of the Equator; latitude lines are all parallel to the Equator 4, 14, 22, 31, 40

latosol   soil found under tropical rainforests 34

leaf litter   dead plant material on the ground surface 34

leisure   for pleasure/recreation 58, 85

levee   ridge of deposited material alongside a channel; results from floods; artificial levees can be made as flood defences 63, 64

LIC   lower income country   8, 11, 16, 74–75, 86–87, 88–89, 102–103, 104–105, 106, 109, 114

limitations   disadvantages or causes of potential errors or the collection of unreliable or unrepresentative data 117

linear   pattern or distribution is linear if there is a concentration of something in a line or ribbon, e.g. a linear settlement pattern 4

load   material carried by an agent of erosion and transport, e.g. river, glacier 62

longitude   longitude lines provide a measure of how far places are east or west from the Prime Meridian (sometimes called the Greenwich Meridian); longitude lines run between the North and South Poles at right angles to the latitude lines 4

longshore drift   movement of material along the shore due to wave action 56–57, 58

long-term aid   providing help for education and skills for development 89

low pressure   at times of low pressure the air is usually rising 14–15, 23, 30

lyme grass   salt-loving plant; one of the first to grow on sand dunes 57, 59

management   the forecasting, planning, organising and decision-making used to improve people's lives   6, 16–17, 19, 35, 39, 59, 64, 98, 106, 108, 119

mantle   the region within the Earth between the core and the crust 8–9

maritime   related to the sea 18

marram grass   tough, long rooted, wide-bladed grass that lives in sand and holds the grains together 57, 59

mass movement   the downhill movement of weathered material under the influence of gravity 49, 53, 54–55

mean   the sum of the quantities divided by the number of quantities 7

meander   winding curve or bend in a river 63

medial (moraine)   merging of two lateral moraines 51

median   the middle value in a ranked set of data   7

mega-city   a city with over 10 million inhabitants 75

meltwater   water resulting from melting snow and ice 50–51, 53

metamorphic   the transformation of rock types as a result of heat and/or pressure at great depths 48

mid-latitudes   the temperate zones between the tropics and polar regions 14, 18

millionaire city   a city with over a million inhabitants 74

mitigation   measures to reduce the causes of climate change, e.g. cutting greenhouse gas emissions; eliminating long-term risk to human life and property; international targets to reduce greenhouse gas emissions; carbon capture 22, 81

modal class   the class with the highest frequency 7

mode   the value that occurs most frequently in a set of data   7

moraine   material carried and deposited by ice 51, 52–53

mouth   where a river ends; usually meeting the sea 59, 60

multipurpose dam   a man-made water control structure used to generate hydroelectric power and for other purposes such as water supply and recreation 101

natural resources   materials occurring in nature that can be used by humans, e.g. oil, wood 87, 100–101

navigation   route for water transport, e.g. barges, boats 64

NEE   newly emerging economy 75, 76, 86

nitrates   agricultural fertilisers 38, 64

North Atlantic Drift   a powerful warm ocean current 18

northing line   one of the horizontal lines crossing an OS map from one side to the other; they are called northings as the numbers increase to the north 4

notch   small indentation in a cliff face at high tide mark; results from wave action 55

nucleated   a pattern or distribution is nucleated if there is a concentration of something in a tightly-confined cluster, e.g. a nucleated settlement pattern 4

nutrient cycle   how the minerals that provide energy and sustenance to living organisms circulate around an ecosystem; minerals are moved between the atmosphere, biomass, litter layer and soil 28, 38

orientation   direction, using compass points or bearings 5

overgrazing   too many animals allowed to feed on a piece of land; leads to loss of plant species and lowering of quality of farmland 53

overland flow   water moving over the surface towards the channel 60-61, 65

oxbow lake   formed when a river meander is cut off 63

pastoral   farming aimed at producing livestock rather than crops 98

percentile   each of the 100 equal groups into which a population can be divided according to the distribution of values of a particular variable 7

peripheral   describes places which are distant from the centre or core of activity 4

permafrost   permanently frozen layer below the Earth's surface 21, 22, 40

permeable   a rock or surface that allows water to pass through 6, 48, 60

physical water scarcity   a situation where the water is not abundant enough to meet all demands for the population 106

plate margin   the boundary between two tectonic plates   8, 10, 12–13

plateau   flat, elevated land rising sharply above the area surrounding it on at least one side 5, 63

plucking   erosion process; weakened rock removed from bedrock by a glacier 50

plunging   tall, powerful, destructive waves that crash on to a beach and take material out to sea 54, 57

polar   regions near the North Pole or South Pole 14, 29, 31, 40–41, 101, 111

political map   shows boundaries linked to areas of governance such as countries, states and counties, and the location of major cities 4

population pyramid   a diagram that shows how the population is made up by gender and age   7, 83

porous   rocks that hold water 48

prevailing wind   a wind from the direction that is most usual at a particular place or season 18

primary data   data that you and your fellow students have collected 117

primary effects   the effects that occur immediately after a natural disaster happens, e.g. the ground shakes 10–11, 12–13, 16–17

pristine   immaculately clean or has never been used 101

problem-solving   thinking about the possible solutions to a problem and deciding which course of action to take 118

producer   an organism, e.g. a tree, that makes its own food via photosynthesis 28–29

proximal   near the point of origin; of a spit: joined to the land 57

proxy measures   indirect ways of measuring variables, such as tree ring growth as an indicator of temperature in past years 20

pull factors   positive factors attracting people to live in a place 75

push factors   negative factors making people want to leave a place 75

pyramidal peak   sharp-pointed, frost-shattered mountain top with corries on at least three sides 51, 52

pyroclastic flow   a high temperature avalanche of gas, ash, cinder and rock that rushes down the slopes of a volcano at speeds of up to 450 mph 12–13

qualitative data   non-numerical data which might involve subjective judgements, e.g. field sketches, photographs, video, quotes or opinion 117

quantitative data   numerical data or statistics, e.g. width and depth of a river channel; an environmental quality survey   7, 117

quartiles   dividing a ranked set of data into four equal groupings   7

Quaternary period   the last 2.6 million years in which there have been several glacials 20

radial   pattern spread out in a linear way from a central point, like the spokes on a bicycle wheel or rays of the Sun, e.g. rivers or glaciers flowing in all directions from high ground; transport links from a hub 4, 6

rain shadow   a dry area on the leeward side of a hilly/mountainous area 18

range   the varying possible outcomes which climate models predict for future temperature, rainfall, etc. 21

recreation   leisure-time activity for pleasure 39, 64

regeneration   reviving old run-down urban areas by either improving what is there or clearing it and rebuilding 79

relief   the difference in elevation in an area; shape of the land 4, 6, 32, 60

relief map   shows the height and shape of the land (topography) using symbols such as contours or layer shading 4

relief rainfall   occurs when moist air rises over a physical barrier, e.g. mountains 18

replacement (of a slope)   a slope changes shape as weathered material stays at the base 49

reprofiling   change the face/front of a slope; could be to make more or less steep, higher or lower, than originally 59

resilience   capacity to cope with a hazard such as a tropical storm, often dependent on how effectively the 3 Ps are implemented 16–17

resistant   tough; able to withstand weathering and/or erosion 54–55

retreat (of a slope)   to wear away backwards but keep the same shape 49

ribbon lake   a long narrow lake formed by a glacier 52

Richter scale   a measure of the energy released by an earthquake 10–11

ridge   mountains forming a continuous elevated crest   5, 52, 54

roche moutonnée   ice-carved rock with a smooth upstream side and rugged downstream side 52

rock fall   mass movement; pieces of weathered rock become detached from the face of a cliff and travel under gravity to the base of the cliff 54

salinisation   the increase in the salt content of water 37

saltation   bouncing or jumping movement of particles along a surface 56–57, 62

salt crystal growth   salt from seawater dries in cracks, joints, etc.; dries and expands on re-wetting, putting pressure on surrounding rocks 48, 54

saltmarsh   area of deposition of fine material that is tidal; adapted plants grow and stabilise the area 57, 59

sampling techniques   data is collected from a small part of the population or a small number of sites and used to inform what the whole picture is like; there are three main types of sampling: random; systematic; stratified 117

sand dune   accumulation of sand in hill form at the back of beaches along low-lying coastlines 57

saturated   no more water can be held; if a permeable material becomes saturated, it becomes impermeable 49, 53, 61

scale   ratio which compares a measurement on a map between places to the actual distance between these places in reality; for example, the most commonly used OS maps use a scale of 1 : 50 000 and 1 : 25 000; scale is also used when referring to places or areas of different sizes, e.g. local, regional, national, international, global scale 5, 118

scenarios   different possible outcomes for climate change, depending on how effectively the anthropogenic causes are managed 21

scree   frost-shattered rocks scattered over a surface or creating a slope 50, 53

sea couch   salt-loving plant; one of the first to grow on sand dunes 59

secondary data   data collected by another person, group or organisation which you have accessed via articles, websites, books, etc. 117

secondary effects   the effects that occur subsequent to the primary effects, such as gas mains rupturing 10–11, 12–13, 16–17

sedimentary rocks that have been produced from layers of sediment, usually at the bottom of the sea, e.g. limestone **48**

settlement a place where people establish a community **4, 32–33, 35, 37, 41, 58–59, 60, 64, 74, 76–77, 78, 81**

shield volcano a wide-based, gentle-sided volcano formed from free-flowing magma **12**

shifting cultivation ground is cultivated and then abandoned to allow its fertility to be restored **35**

short-term aid immediate relief after an emergency **89**

sinuous curving movements in a river's flow **63**

slide mass movement; downslope movement of material that holds together until it reaches the base of a slope when it breaks up **49, 54**

slip-off slope area of deposition at the inside bend of a meander **63**

slump mass movement; downslope movement that follows a curved or rotating path; usually on weaker rocks **49, 54**

soft engineering range of measures used to manage rivers and coasts which do not use materials such as concrete, stone or steel, e.g. river restoration, floodplain zoning; they allow natural processes to occur and don't alter the natural environment; often less expensive than hard engineering and considered to be more economically, socially and environmentally sustainable **6, 58, 65**

solifluction slow mass movement of wetter material **49**

solution dissolving action in water **54, 63**

source beginning of a river **60**

sparse a pattern or distribution is sparse if it shows a low density and/or is very dispersed in an area, e.g. sparse drainage pattern; sparse population **4**

spilling low, gentle, constructive waves that spill on to a beach and push material onshore **54**

spit long protrusion of deposited material extending from the coast into the sea **57**

spot heights points of known height shown using a dot with a number beside it, usually in metres above sea level; they appear on a map, but not on the landscape **5**

spur a piece of land jutting into a river or stream **5, 52, 62**

stack isolated rock pillar in the sea **55**

stakeholders people or groups with a real interest in an issue or in the development of a decision or policy **118–119**

storm surge a rising of the sea as a result of wind and atmospheric pressure changes associated with a storm **15, 16–17, 22**

stratification a system or formation of layers **34**

striations scratches cut into bedrock by glacial abrasion **50**

stump eroded stack usually only visible at low tide **55**

subdued description of a river which responds very slowly to rainstorms **61**

sub-glacial (moraine) debris under the ice **51**

subsistence farming crops produced solely for the farmer and family **35, 87**

supra-glacial (moraine) debris on top of the ice **51**

surging wave which pushes material up a beach, making it steeper **54**

suspended load particles carried along within the body of water (river or sea) **62**

suspension material carried within the body of the water; not floating **56**

sustainable meeting the needs of the present population without compromising the needs of future generations **35, 39, 81, 87, 104–105, 108–109, 114–115, 119**

Sustainable Development Goals list of goals that the United Nations would like countries to meet **87**

swash forward wave motion **54, 56**

symmetrical similar gradient on both sides **6**

synoptic thinking the application of knowledge, understanding and skills from all of your geographical learning to a new situation **118**

teleconnections extreme or unusual weather events which take place a very long way from the Pacific and are thought to be indirectly linked to ENSO **23**

temperate descriptive of something, such as an ecosystem, that can be found above 45° north and south of the Arctic Circle **14, 18, 29, 31, 39**

terminal (moraine) crescent-shaped deposit at the furthest reach of a glacier **51, 52**

thermal expansion growth in volume of substances as they become warmer, including seawater, which contributes to sea level rise **22**

throughflow lateral movement of water below the surface (through the soil towards the river) **61**

till clay-containing rock fragments deposited from under a glacier **51, 52**

transnational corporation (TNC) enterprise operating in at least two countries **88–89**

traction rolling and sliding of larger, heavier particles along a river bed or in the sea **56, 62**

transpiration water evaporates from plants into the atmosphere **34, 36, 39, 60**

transportation material being carried away; in rivers, by solution, suspension, traction and saltation **50–51, 56–57, 58, 62, 64**

triangulation pillars points of very precise known height shown using a dot with a number beside it, usually in metres above sea level; these appear on a map as a triangular symbol, and are marked on the landscape with a concrete pillar **5**

tributary any river flowing into another, bigger river **52, 60**

tropical descriptive of something, such as an ecosystem, that can be found on or near to the Tropics of Cancer and Capricorn **28–29, 31, 34–35, 38, 101**

tropical storm low-pressure systems with distinct structure and features formed over warm ocean waters in low latitudes **4, 14–15, 16–17, 87**

trough glaciated valley; also known as a U-shaped valley **51, 52**

truncated spur former interlocking spur from a river valley which has been cut away by glacial erosion **52**

tsunami a large wave caused by an earthquake, volcanic eruption or coastal landslide **10–11**

tundra vast treeless zone in the Arctic regions of the world **29, 31, 40**

upstream in the opposite direction to the way the river is flowing **52**

urbanisation an increase in the number of people living in towns and cities compared with rurally **64, 74**

valley between areas of higher land, often with a river running through **5, 6, 8, 48, 51, 52–53, 60–61, 62–63, 64**

warm front warm air is advancing and rising up over cold air, because warm air is less dense than cold air; warm air is therefore replacing cooler air at the surface **14**

waterborne disease caused by coming into contact with an infected water source **106, 109**

water conservation beneficial reduction of water usage **108**

water deficit not enough rainfall, leaving a water shortage **19**

watershed the dividing line between two drainage basins **60**

water stress lacking water **19**

water surplus having more water than needed **19, 106**

water transfer scheme moving water from an abundant river basin to one where water is less available **99, 108**

wave cut platform gently shelving area of solid rock stretching out to sea from the cliff front **55, 56**

weathering the breakdown of rock by mechanical, chemical or biological processes **48–49, 50, 52–53, 54–55, 57**

wetted perimeter length of channel bed in contact with water **60–61**

wilderness areas untouched by human influence **41, 101**

wind patterns the semi-predictable movement of masses of air around the world, e.g. the trade winds **30**

yield the measure of crops produced, often in kilograms per hectare **103, 104**

Collins

GCSE Revision
Geography

GCSE Workbook

Janet Hutson, Dan Major, Paul Berry,
Brendan Conway and Robert Morris

Revision Tips

Rethink Revision

Have you ever taken part in a quiz and thought *'I know this!'* but, despite frantically racking your brain, you just couldn't come up with the answer?

It's very frustrating when this happens but, in a fun situation, it doesn't really matter. However, in your GCSE exams, it will be essential that you can recall the relevant information quickly when you need to.

Most students think that revision is about making sure you **know** stuff. Of course, this is important, but it is also about becoming confident that you can **retain** that *stuff* over time and **recall** it quickly when needed.

Revision That Really Works

Experts have discovered that there are two techniques that help with all of these things and consistently produce better results in exams compared to other revision techniques.

Applying these techniques to your GCSE revision will ensure you get better results in your exams and will have all the relevant knowledge at your fingertips when you start studying for further qualifications, like AS and A Levels, or begin work.

It really isn't rocket science either – you simply need to:

- **test yourself** on each topic as many times as possible
- **leave a gap** between the test sessions.

Three Essential Revision Tips

1. **Use Your Time Wisely**
   - Allow yourself plenty of time.
   - Try to start revising at least six months before your exams – it's more effective and less stressful.
   - Your revision time is precious so use it wisely – using the techniques described on this page will ensure you revise effectively and efficiently and get the best results.
   - Don't waste time re-reading the same information over and over again – it's time-consuming and not effective!

2. **Make a Plan**
   - Identify all the topics you need to revise (this All-in-One Revision & Practice book will help you).
   - Plan at least five sessions for each topic.
   - One hour should be ample time to test yourself on the key ideas for a topic.
   - Spread out the practice sessions for each topic – the optimum time to leave between each session is about one month but, if this isn't possible, just make the gaps as big as realistically possible.

3. **Test Yourself**
   - Methods for testing yourself include: quizzes, practice questions, flashcards, past papers, explaining a topic to someone else, etc.
   - This All-in-One Revision & Practice book provides seven practice opportunities per topic.
   - Don't worry if you get an answer wrong – provided you check what the correct answer is, you are more likely to get the same or similar questions right in future!

Visit our website to download Your free flashcards, for more information about the benefits of these techniques, and for further guidance on how to plan ahead and make them work for you.

www.collins.co.uk/collinsGCSErevision

Contents

Natural Hazards

Refer to this map for questions 1–3.

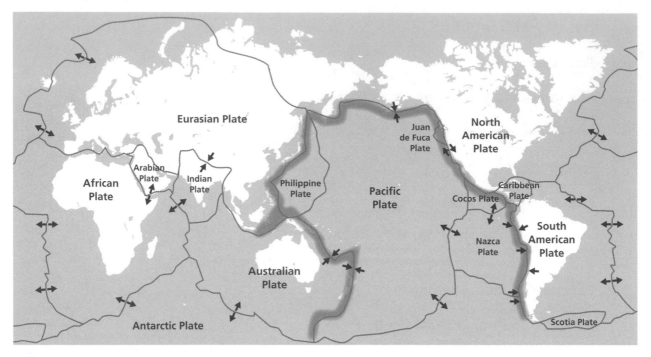

1 What is the common name given to the highly active area of tectonic activity highlighted on the map? [1]

_____

2 Use the map to help you name two plates that form a constructive margin. [2]

_____

_____

3 Use the map to help you name two plates that form a destructive margin. [2]

_____

_____

4 Name an example of a major earthquake and state where and when it happened. [2]

_____

_____

5 What is the difference between the focus and the epicentre of an earthquake? [2]

_____

_____

6 Describe ways of predicting when and where an earthquake might happen. [6]

7 Describe what you think might be contained in an earthquake survival kit for a typical householder. [6]

8 Giving a named example, describe the effects of a tsunami. [6]

Natural Hazards

9 With reference to the Merapi volcanic eruption of 2010, state whether the following statements are true or false: [5]

a) It occurred at a destructive plate margin. ....

b) It produced a giant ash plume. ....

c) It resulted in a major pyroclastic flow. ....

d) It produced a serious lahar. ....

e) It occurred beneath an ice cap. ....

10 What name is given to the effects of ENSO, which are thought to happen a long way from the Pacific Ocean (e.g. more snow in the Rocky Mountains of the USA and Canada; milder, wetter winters in the UK)? [1]

....

11 a) Describe the pattern of tropical storm tracks in the map below. [4]

Tropical Storm Tracks Over the Last 70 Years

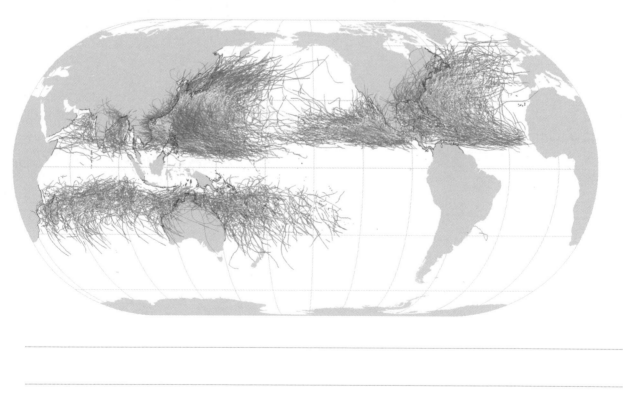

....

....

....

....

b) Describe and explain the relationship between the pattern of tropical storm tracks and average sea surface temperatures. [4]

Average Sea Surface Temperatures

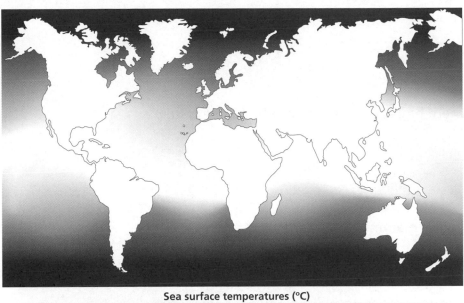

Sea surface temperatures (°C)

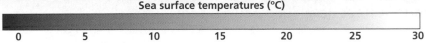

0   5   10   15   20   25   30

....

....

....

....

12 The chart shows changes in carbon dioxide from the Mauna Loa Observatory in Hawaii.

Why is this location considered to be a useful place to measure average carbon dioxide in the atmosphere? [2]

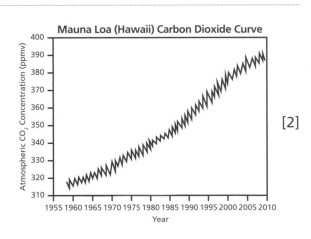

....

....

Total Marks .... / 43

Ecosystems

1 Describe one example of how human action can disrupt the balance within a particular ecosystem. [3]

....

....

....

2 Give two places where high pressure cells can be found in July. [2]

....

....

3 Name one water-loving mammal found in the UK. [1]

....

4 Why have some rainforest trees developed buttress roots? [2]

....

....

5 What problem is caused by over-irrigation? [1]

....

6 How does global warming affect the stability of deciduous forests? [1]

....

7 What term describes the niche in the tourism industry for travel to new, undiscovered and potentially hazardous environments? [1]

....

8 Name three of the UK's main ecosystem types. [3]

_____

_____

_____

9 What word describes how forests develop in clear 'layers'? [1]

_____

10 Outline the causes of deforestation. [4]

_____

_____

_____

_____

11 Describe the distribution of the world's hot deserts. [3]

_____

_____

_____

12 Outline an adaptation of one desert species. [2]

_____

_____

Total Marks _____ / 24

Physical Landscapes in the UK

1 Describe the weathering of granite into a tor. [4]

....

....

....

....

2 Look at the diagrams.

Which glacier, **A** or **B**, would be most effective at erosion and why? [3]

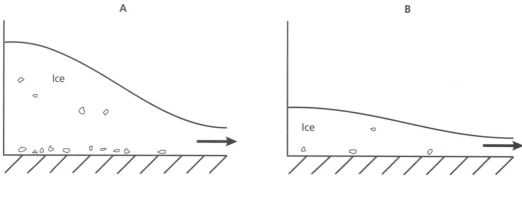

....

....

....

3 Why are both erosion and deposition evident in lowland areas but only erosion in highland areas? [3]

....

....

....

4 How do groynes interfere with the action of longshore drift? [2]

....

....

5 The diagram below shows summer and winter profiles for the same beach.

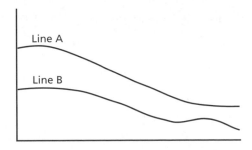

Line A

Line B

a) Which line is the summer profile, and which the winter? [1]

b) Explain why the winter profile is the shape that it is. [3]

6 Compare the action of spilling and surging waves. [2]

7 What is meant by a concordant coastline? [1]

8 A section of coastline containing headlands and bays is likely to become straighter over time. Explain why this will happen. [4]

9 This photograph shows a sign next to Godrevy beach in Cornwall.

a) What does the sign mean by 'failure' of the cliff? [2]

_____

_____

_____

> Cliff subject to ongoing
> failure particularly during
> and following wet and
> stormy weather.
> Cliff edges often undercut.
> Keep well back.
>
> National
> Trust   Godrevy

b) How will wet weather change the cliff to make 'failure'
more likely? [2]

_____

_____

c) How will stormy weather make 'failure' more likely? [2]

_____

_____

10 Describe the action and results of abrasion on a cliff. [4]

_____

_____

_____

_____

11 a) For a named section of coastline in the UK that you have studied, explain why coastal
protection was considered to be necessary. [2]

_____

_____

b) Describe one of the methods of coastal protection used on that same section of coastline. [2]

_____

_____

12 Locate, name and describe **one** feature caused by coastal erosion and **one** feature caused by coastal deposition. [6]

Location of place/area:

Erosion feature:

Deposition feature:

13 Explain some of the effects on a drainage basin of building on a field. [4]

14 Look at this photograph taken near Hackfall on the River Ure, North Yorkshire.

What does the existence of the structure in the foreground suggest? [3]

Total Marks .... / 50

The Human Environment

1 What is a 'millionaire city'?   [1]

....

2 What is a 'mega-city'?   [1]

....

3 Identify three reasons why world birth rates are falling.   [3]

....

....

....

4 Which of the following statements are true for a higher income country (HIC)?
Tick the correct answers.   [2]

A Low birth rates

B High death rates

C High elderly population

D High infant mortality

E Rapidly increasing population

5 Why do people choose to leave large cities to live in smaller settlements nearby?   [3]

....

....

....

6 How did local residents benefit from the London Olympics in 2012?   [4]

....

....

....

....

7 Describe the living conditions in a favela. [4]

_____

_____

_____

_____

8 Name three examples of new towns in the UK. [3]

_____

_____

_____

9 Label the following land use zones in a theoretical UK city. [6]

A _____

B _____

C _____

D _____

E _____

F _____

10 What changes can be made to modern cities to make them 'carbon neutral'? [4]

_____

_____

_____

_____

The Human Environment

11 Describe an example of diversification in modern agriculture. [2]

12 Give three reasons why the UK has a declining death rate. [3]

13 Name two jobs that make up the primary sector. [2]

14 Name two jobs that make up the quaternary sector. [2]

15 Draw a sketch of a population pyramid for a country in stage 2 of the Demographic Transition Model (DTM). Add labels to explain the main features. [6]

16 What threats do our oceans face in the modern world? [5]

_____

_____

_____

_____

_____

17 Draw lines to match each type of aid to the correct definition. [4]

Type of Aid	Definition
Long-term aid	Aid that is provided by a group of countries
Voluntary aid	Aid that solves an immediate crisis
Multi-lateral aid	Aid that solves a problem so that it does not occur again
Short-term aid	Aid that comes with conditions
Tied aid	Aid that comes from charities

Total Marks _____ / 55

Resources

1 What is meant by a multipurpose dam? [2]

2 How has the way that the UK uses different energy sources changed in the last 50 years? [4]

3 Why might the use of different types of renewable energy not be suitable for all countries? [3]

4 Why is increasing food security in lower income countries a difficult challenge? [4]

5 Explain what is meant by 'organic farming'. [3]

6 What is 'groundwater'? [1]

7 Define the term 'biofuel'. [1]

8 Study the figure below. Suggest reasons why some regions of the world have water scarcity. [4]

Fresh Water in the World
Access to renewable water sources (m³ per person, per year)

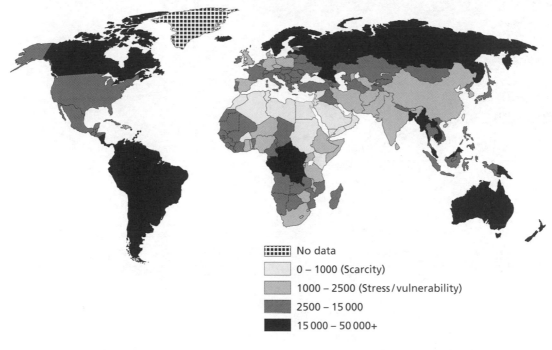

▦	No data
	0 – 1000 (Scarcity)
	1000 – 2500 (Stress / vulnerability)
	2500 – 15 000
	15 000 – 50 000+

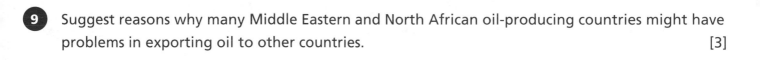

9 Suggest reasons why many Middle Eastern and North African oil-producing countries might have problems in exporting oil to other countries. [3]

10 What special condition is needed for the generation of geothermal energy? [1]

11 Name a country that relies heavily on nuclear power. [1]

Total Marks _____ / 27

Geographical Applications

1 For your geographical enquiry, describe and explain patterns in your data and any anomalies which did not correspond to the main patterns. [9 + 3 SPaG]

_____

_____

_____

_____

_____

_____

_____

_____

_____

_____

Total Marks _____ / 12

Collins

GCSE
GEOGRAPHY

Paper 1 Physical Geography

Time allowed: 1 hour 30 minutes

Materials

For this paper you must have:

- a pencil
- a ruler.

Instructions

- Use black ink or a black ball-point pen.
- Answer all questions for your board and specification.
- Cross through any work you do not want to be marked.

Information

- The marks for questions are shown in square brackets **[ ]**.
- The total number of marks available for this paper is 86.
- Spelling, punctuation, grammar and specialist terminology will be assessed in Question **1A f)** or Question **1B f)** depending on your board and specification.

Name: _____

Question 1

Question 1A is for AQA, Edexcel B, Eduqas A and OCR B

Question 1B is for Edexcel A, Eduqas B and OCR A

Question 1A Earthquakes and volcanoes

Study **Figure 1**, a map showing the major tectonic plates and their direction of movement.

Figure 1

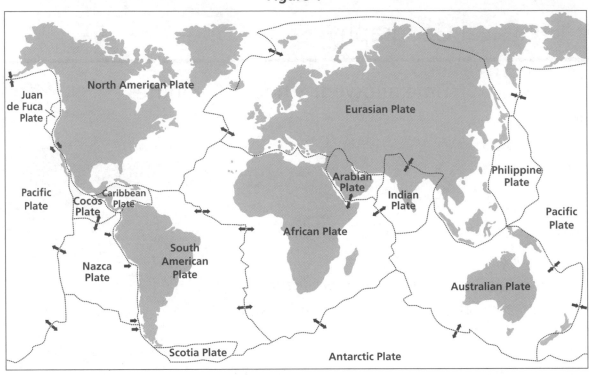

a) Which **two** words could be used to complete the following statement?

The plate boundary shown between the Nazca Plate and the Pacific Plate is:

Tick **two** options only.   [2]

A   convergent

B   divergent

C   constructive

D   destructive

E   collision

F   conservative

b) Complete the following sentence.

An example of a conservative plate boundary is the boundary between the

_____ plate and the _____ plate.   [1]

Study **Figure 2**, a photograph taken in Nepal following a magnitude 7.8 earthquake in April, 2015.

Figure 2

c) Explain what is meant by the term 'magnitude 7.8'. [2]

d) Identify the impact on the landscape shown in **Figure 2**. [1]

e) Suggest **two** ways in which the impact shown in **Figure 2** could affect people living in the area. [4]

Question 1A continues on the next page

Study **Figure 3**, an extract about Mount Etna, a volcano on the Italian island of Sicily, and **Figure 4**, a photograph showing a house in Sicily following a recent eruption.

Figure 3   **Figure 4**

Etna produces pyroclastic flows, ash falls and mudflows. However, the most hazardous type of activity is lava flows, in particular to Catania, a city of 300,000 people and the second largest in Sicily. Lava flows do not usually move fast enough to present danger to the population, but they can cover large areas and destroy crops and buildings. Evacuation of the area around Etna could be done only with great difficulty.

In this question, 3 of the marks awarded will be for spelling, punctuation and grammar and for your use of specialist terminology.

f)   Describe and explain the possible hazards and benefits of living near an active volcano. Use **Figure 3** and **Figure 4** and an example you have studied.   **[12]**

Total marks for Question 1A = 22

Question 1B Weather hazards

Study **Figure 5**, a chart showing an area of high pressure over the UK in July, 2013.

Figure 5

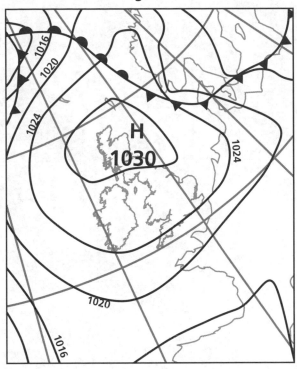

a)   Name the type of chart shown in **Figure 5**.   [1]

_____

b)   State the highest pressure shown in **Figure 5**. Include the unit of measurement.   [1]

_____

c)   Describe how the pattern of isobars shown in **Figure 5** indicates wind strength.   [2]

_____

_____

Question 1B continues on the next page

Study **Figure 6**, a satellite image of the UK taken at the same time as **Figure 5**.

Figure 6

d)   Describe the cloud coverage shown in **Figure 6**.   [1]

Study **Figure 7**, a description of UK weather conditions in July, 2013.

Figure 7

2013 Heatwave

Between 6 July and 24 July 2013, many places in the UK experienced temperatures above 28°C. From the 13th to the 19th, there were temperatures of over 30°C with a high of 33.5°C recorded in Northolt, Greater London. Night time temperatures remained above 16°C.

In general, it was slightly cooler at the coast.

The heatwave ended with storms on 22 and 23 July. A total of 35.6 mm rain fell in one hour in Nottinghamshire.

e)   State why coastal areas would have been cooler than inland.   [1]

In this question, 3 of the marks awarded will be for spelling, punctuation and grammar and for your use of specialist terminology.

f) Describe and explain the negative impacts that continuous high pressure could have on the population of the UK. [12]

g) The heatwave ended with a storm. What type of rainfall is associated with this type of storm? [1]

h) There was flooding in many areas of the UK.
Explain how the heatwave would have contributed to this flooding. [2]

i) Name a weather hazard associated with high pressure in winter. [1]

Total marks for Question 1B = 22

Question 2

Question **2A** is for **AQA, Edexcel A, Edexcel B, Eduqas A** and **Eduqas B**

Question **2B** is for **OCR A** and **OCR B**

Question 2A Ecosystems and rainforests

Study **Figure 8**, a world map showing some large-scale ecosystems (biomes).

Figure 8

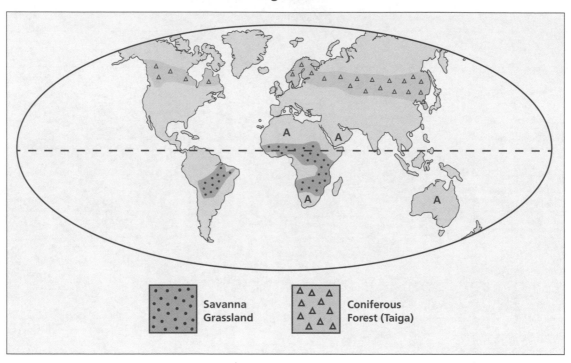

a) Using **Figure 8** and your own knowledge, choose the correct words to complete the following sentences.

Circle the correct answer from those given in bold.

Coniferous forests are found **mainly / entirely** in the Northern Hemisphere.

They are located to the **north / south** of tundra regions.

There is **some / no** savanna grassland in Asia.   [3]

b) State which type of large-scale ecosystem you would expect to find in the areas labelled **A** on **Figure 8**.   [1]

Study **Figure 9**, a simple food chain.

Figure 9

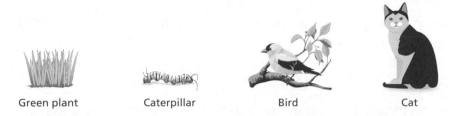

| Green plant | Caterpillar | Bird | Cat |

c) Add arrows to **Figure 9** to show the direction of energy flow. **[1]**

d) State which of the four organisms shown in **Figure 9** is classed as a producer. **[1]**

Study **Figure 10**, showing the nutrient cycle in a tropical rainforest.

Figure 10

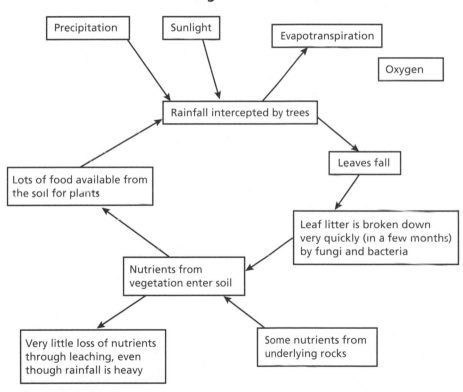

e) Draw an arrow on **Figure 10** to connect oxygen to the cycle.

The direction of your arrow should show whether oxygen is an input or an output. **[1]**

Question 2A continues on the next page

f) Define the term 'evapotranspiration'. [1]

g) State the reason why leaf litter is broken down much quicker in the rainforest than in woodlands in the UK. [1]

h) Name the process that enables nutrients to be released from the underlying rocks. [1]

i) Explain how plants and animals in the rainforest are adapted to the weather and climate. [6]

Total marks for Question 2A = 16

Question 2B Physical Landscapes in the UK

a)   Define the term 'weathering'.   [1]

....

Study **Figure 11**, a photograph showing an upland area of the UK.

Figure 11

b)   There are scree slopes visible in **Figure 11**.

Describe the weathering process that forms scree slopes.   [3]

....

....

....

c)   State what evidence there is in **Figure 11** to suggest that the scree slopes are no longer

being actively formed.   [1]

....

d)   Suggest an area in the UK where the photograph in **Figure 11** could have been taken.   [1]

....

Question 2B continues on the next page

Study **Figure 12**, a photograph showing a different upland area of the UK.

Figure 12

e)   Give one piece of evidence to suggest that the area in **Figure 12** is at a lower level than the area in **Figure 11**.   **[1]**

_____

Study **Figure 13**, a map and table showing the location and climate data for four places in the UK.

Figure 13

	A	B	C	D
Total Rainfall (mm)	1017	1329	619	2612
Number of Days of Rain	154	171	117	200
July Mean Temperature (°C)	16.4	14.5	16.5	14.5

f) Draw a line from each number to a letter, to match the location to the correct data set. **[3]**

1   A

2   B

3   C

4   D

g) Explain why rainfall varies across the UK. **[6]**

....

....

....

....

....

....

Total marks for Question 2B = 16

Question 3

Question 3 is for **all** boards and specifications

Answer **one** question only

Question 3   Ecosystems

a)   For a named large-scale ecosystem (biome), discuss the links between its climate, vegetation and animals.

OR

b)   For a named large-scale ecosystem (biome), discuss the threats and attempts to manage them.

[8]

....

....

....

....

....

....

....

....

....

....

....

....

Total marks for Question 3 = 8

Question 4

Question 4 is for all boards and specifications

Question 4   Coastal landscapes in the UK

Study **Figure 14**, a photograph from a beach in Suffolk.

Figure 14

a)   Describe the erosion process that produces the rounded beach material
shown in **Figure 14**.   [2]

_____

_____

Study **Figure 15**, a photograph of Dunwich in Suffolk.

Figure 15

A

b)   Name the process of mass movement that has taken place at **A** in **Figure 15**.   [1]

_____

c)   State a rock type that is often associated with the process illustrated by **Figure 15**.   [1]

_____

Question 4 continues on the next page

Study **Figure 16**, a photograph showing a section of the coastline of the Isle of Anglesey.

Figure 16

d) Name the feature that is labelled **B** on **Figure 16**. [1]

Study **Figure 17**, a 1 : 25 000 map extract of part of the coastline of the Isle of Mull.

e) Using **Figure 17**, match the three coastal features shown in the table below to the correct grid references.

Write the letter of the correct grid reference in the table. [3]

Figure 17

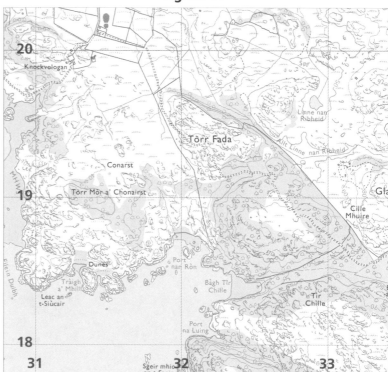

A 318184	B 321195
C 312182	D 324184
E 322182	F 314190

Feature	Grid reference
Headland	
Sandy beach	
Stack	

f) State **two** characteristics of a constructive wave. [2]

Total marks for Question 4 = 10

Question 5

Question 5 is for all boards and specifications

Question 5   River landscapes in the UK

Study **Figure 18**, a photograph showing a river in the Lake District at low flow.

Figure 18

a)   Give **one** feature shown in **Figure 18** which provides evidence that the channel has had a lot more water in it in the past.   [1]

_____

b)   Give **one** feature shown in **Figure 18** which provides evidence that the channel has not had a lot more water in it for a long time.   [1]

_____

Question 5 continues on the next page

Study **Figure 19**, a photograph of Janet's Foss, a waterfall in North Yorkshire.

Figure 19

c) Define the term 'waterfall'. [1]

d) Explain the formation of a waterfall and its associated features. [6]

e) State where on a river meander the water flow is fastest. [1]

Total marks for Question 5 = 10

Question 6

Question 6 is for **all** boards and specifications

Question 6   Climate change

a)   Define the term 'climate'.   [1]

....

b)   One piece of evidence for climate change is provided by the seasons.

With reference to the UK, explain how the seasons provide evidence for climate change.   [2]

....

....

Study **Figure 20**, a graph showing changes to glaciers since 1980.

Figure 20

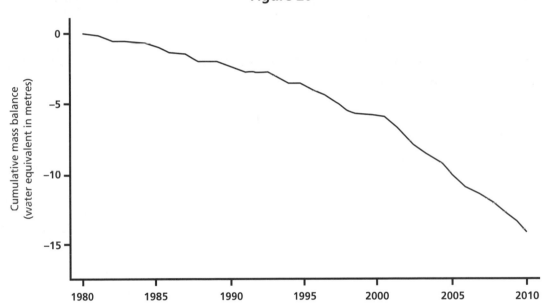

c)   Describe the pattern of glacier reduction shown in **Figure 20**.   [3]

....

....

....

Question 6 continues on the next page

d) Which **two** of the following statements about greenhouse gases are true?

Tick **two** options only. **[2]**

A   Greenhouse gases absorb heat from the Sun. ☐

B   Greenhouse gases make the surface of the Earth cooler. ☐

C   Greenhouse gases are essential to life on Earth. ☐

D   Greenhouse gases absorb heat from the surface of the Earth. ☐

e) Explain how deforestation is contributing to the enhanced greenhouse effect. **[2]**

....

....

f) Give **two** ways in which rainfall in the UK is changing as a result of climate change. **[2]**

....

....

g) Describe the ways in which climate change might make the quality of life worse for some named areas of the UK. **[6]**

....

....

....

....

....

....

h) The climate of the UK has not always been the same.

Describe **one** piece of evidence that shows this statement to be true. **[2]**

....

....

Total marks for Question 6 = 20

END OF QUESTIONS

Collins

GCSE
GEOGRAPHY

Paper 2 The Human Environment

Time allowed: 1 hour 30 minutes

Materials

For this paper you must have:

- a pencil
- a ruler.

Instructions

- Use black ink or a black ball-point pen.
- Answer all questions for your board and specification.
- Cross through any work you do not want to be marked.

Information

- The marks for questions are shown in square brackets [ ].
- The total number of marks available for this paper is 78.
- Spelling, punctuation, grammar and specialist terminology will be assessed in Question **3 b)** or Question **3 c)**.

Name: ....

Question 1

Question **1A** is for **AQA**, **Edexcel B**, **Eduqas B**, **OCR A** and **OCR B**

Question **1B** is for **Edexcel A** and **Eduqas A**

Question 1A Urban issues and challenges

Study **Figure 1**, a graph showing the change in the percentage of the global population living in urban areas over a 50-year period.

Figure 1

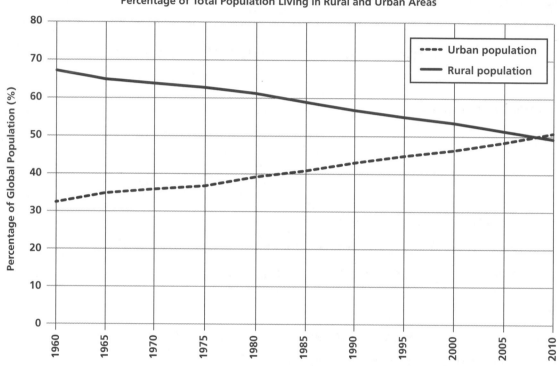

a) Using **Figure 1**, describe **two** changes that occurred to the urban and rural populations between 1960 and 2010. [2]

....

....

b) Give **two** factors that have led to rapid urbanisation in lower income (developing) countries and newly emerging (newly industrialised) countries. [2]

....

....

Study **Figure 2**, a world map showing cities of over five million inhabitants.

Figure 2

Cities of over Five Million Inhabitants

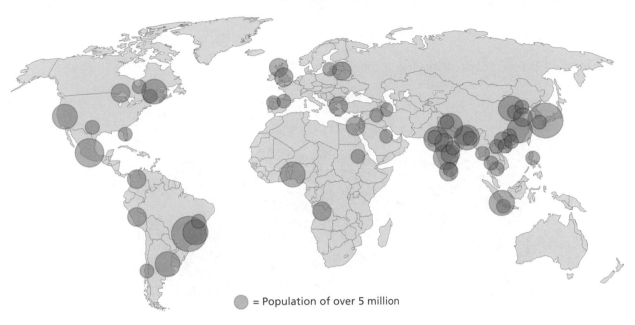

⬤ = Population of over 5 million

c) A mega-city is an urban area with a population of over 10 million.

From **Figure 2**, state which continent has the greatest number of mega-cities. **[1]**

_____

Study **Figure 3**, a photograph showing a slum in Mumbai.

Figure 3

d) Considering the evidence from cities such as Mumbai, explain how growth in cities in lower income (developing) countries or newly emerging countries can create challenges for their populations. **[6]**

_____

_____

_____

_____

_____

_____

Total marks for Question 1A = 11

Question 1B Rural–urban links

a) Give **one** way in which rural and urban areas in lower income (developing) countries differ.   [1]

_____

Study **Figure 4**, a photograph showing a scene from rural Kenya.

Figure 4

b) Using **Figure 4** and your own knowledge, give **two** reasons why a person might want to move away from this area.   [2]

_____

_____

c) Describe what is meant by 'shanty town' or 'squatter settlement'.   [2]

_____

_____

d) Discuss the impact of technological change on service provision in a named rural area in the UK.   [6]

_____

_____

_____

_____

_____

_____

Total marks for Question 1B = 11

Question 2

Question 2A is for AQA, Edexcel A, Edexcel B and Eduqas B

Question 2B is for Eduqas A, OCR A and OCR B

Question 2A Urban development in the UK

Study **Figure 5**, an Ordnance Survey map extract of Oxford.

Figure 5

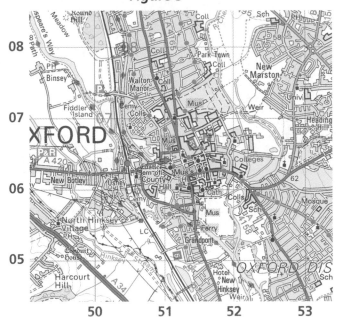

a) Which grid square shows the CBD?

Tick **one** option only.   [1]

A   5106   ☐   B   5005   ☐   C   5207   ☐

D   5205   ☐   E   5105   ☐

b) Explain how a regeneration project that you have studied has helped to improve the
social and economic conditions of its locality.   [6]

_____

_____

_____

_____

_____

_____

Total marks for Question 2A = 7

Question 2B Urban issues and challenges

a) What type of communities arise when resources are used in such a way so as to not cause their depletion? [1]

b) Explain, using a named example, how an out-of-town shopping centre can have both costs and benefits for the local community and area. [6]

Total marks for Question 2B = 7

Question 3

Question **3** is for **all** boards and specifications

Answer part **a)**, plus either **b)** or **c)**

Question 3   Urban issues and challenges

a)   Choose from the list **two** features of sustainable urban living.

Tick **two** options only.   **[2]**

A   Waste recycling   ☐   B   Creating green space   ☐

C   Large-scale road building ☐   D   Landfill sites   ☐

E   Building on greenfield sites ☐

In this question, 3 of the marks awarded will be for spelling, punctuation and grammar and for your use of specialist terminology.

b)   Using a named example, explain how urban transport strategies are being used to reduce traffic congestion.   **[12]**

OR

c)   Using a named example, explain how global cities are connected to the rest of the world.   **[12]**

Total marks for Question 3 = 14

Question 4

Question **4** is for **all** boards and specifications

Question 4   The changing economic world

a)   Define the term 'birth rate'.   **[2]**

_____

_____

b)   Apart from birth rate, give **three** other social measures of development.   **[3]**

_____

_____

_____

c)   Explain the limitations of using economic indicators as a means to measure development.   **[4]**

_____

_____

_____

_____

d)   Give **two** strategies for narrowing the development gap.   **[2]**

_____

_____

Total marks for Question 4 = 11

Question 5

Question **5A** is for **AQA, Edexcel A, Edexcel B, Eduqas A** and **Eduqas B**

Question **5B** is for **OCR A** and **OCR B**

Question 5A Economic development

Study **Figure 6**, a map of Silicon Roundabout in London.

Figure 6

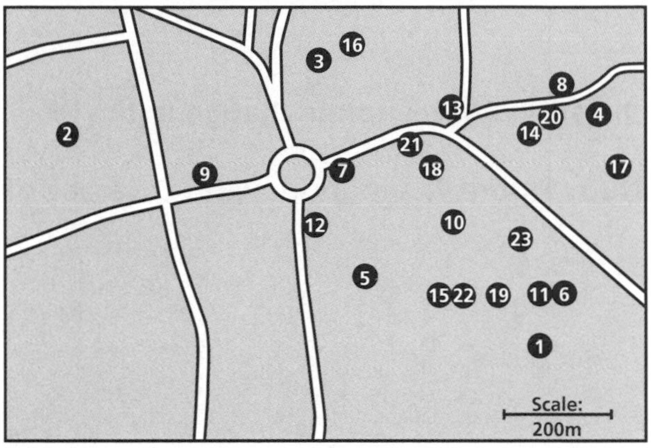

Scale: 200m

a) Using **Figure 6**, measure the direct distance in metres between businesses 5 and 14. [1]

b) Places such as Silicon Roundabout in London and Silicon Fen in Cambridge have seen huge recent growth in high-tech industry. Suggest **three** features that a high-tech company would be looking for when choosing where to locate its facilities. [3]

c) Explain how economic development is improving the quality of life for the population of a place you have studied. [8]

Question 5A continues on the next page

d) Give **two** ways in which modern industrial development in more developed countries, such as the UK, can be made more sustainable. [2]

Total marks for Question 5A = 14

Question 5B Economic change in the UK

Study **Figure 7**, a map showing coverage of superfast broadband across the UK.

Figure 7

Superfast Broadband Coverage

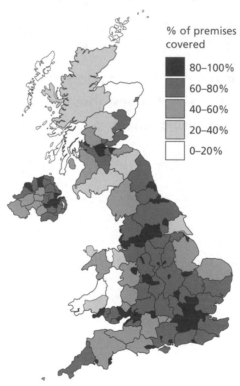

% of premises covered

- 80–100%
- 60–80%
- 40–60%
- 20–40%
- 0–20%

a) Using **Figure 7**, name one part of the UK that has 0–20% of premises covered by superfast broadband. [1]

b) Suggest **three** reasons for the geographical variations in access to superfast broadband. [3]

Study **Figure 8**, the Demographic Transition Model.

Figure 8

Demographic Transition Model

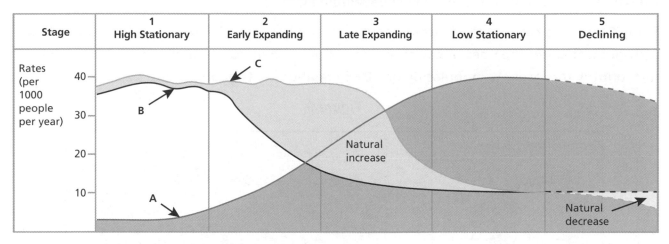

c) Label the three lines on **Figure 8** appropriately. [2]

A ....

B ....

C ....

d) Explain how the UK can respond to the challenges of an ageing population. [8]

....

....

....

....

....

....

....

....

Total marks for Question 5B = 14

Question 6

Question 6 is for all boards and specifications

Question 6   The challenges of uneven development

Study **Figure 9**, a cartogram of world population and life Gross Domestic Product (GDP). This cartogram changes the sizes of countries in relation to their population and GDP. The larger the country, the larger its population or GDP in relation to its geographical size.

Figure 9

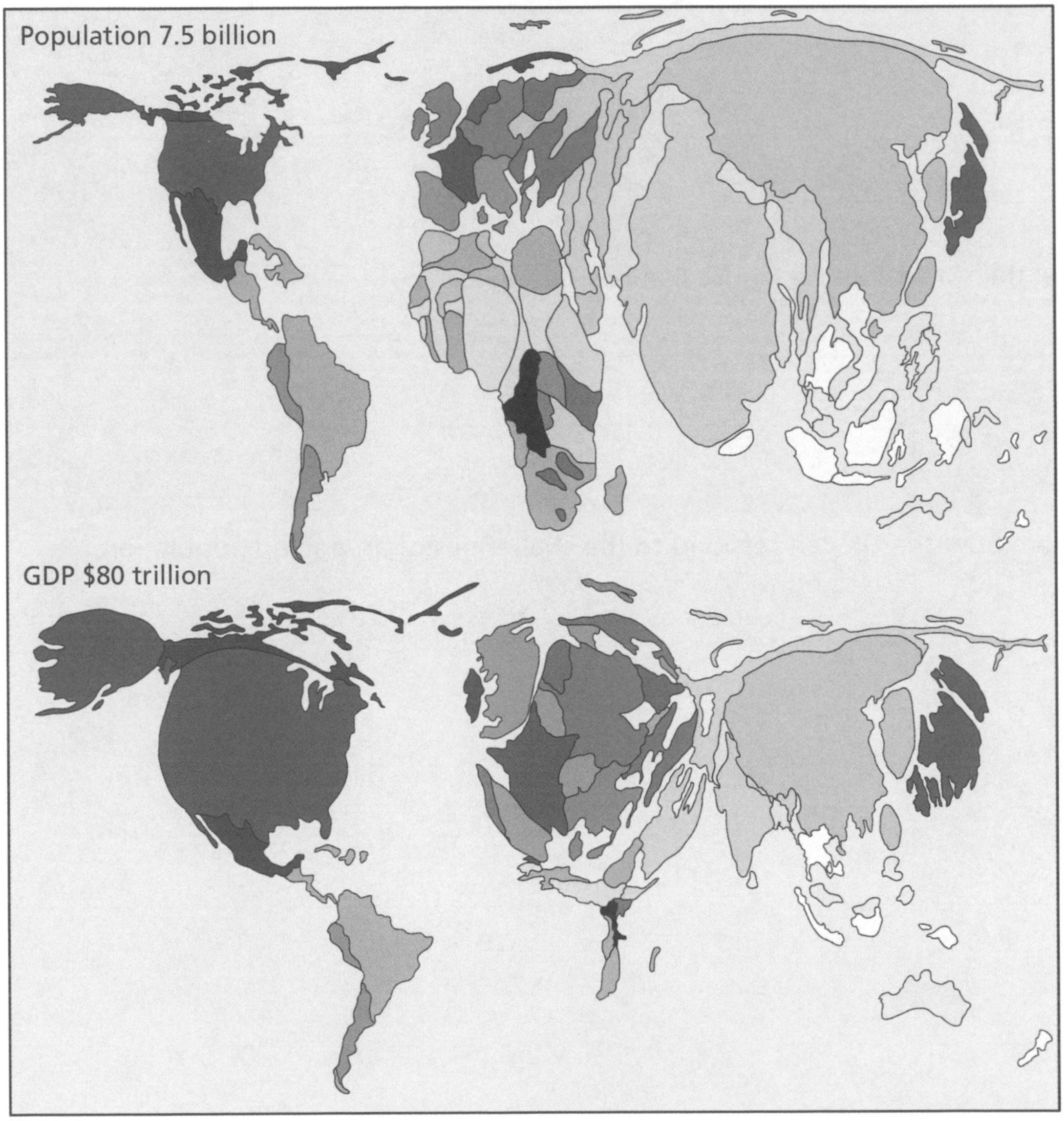

Population 7.5 billion

GDP $80 trillion

a) Using **Figure 9**, name **one** part of the world that shows a negative correlation/link between population and GDP. [1]

....

b) Apart from GDP, give **two** other ways of measuring a country's level of development. [2]

....

....

c) Outline the causes of uneven development. [4]

....

....

....

....

d) Describe the ways in which more developed countries, such as the USA, can give aid to less developed countries, such as Malawi. [4]

....

....

....

....

Question 6 continues on the next page

Study **Figure 10**, a cartoon.

Figure 10

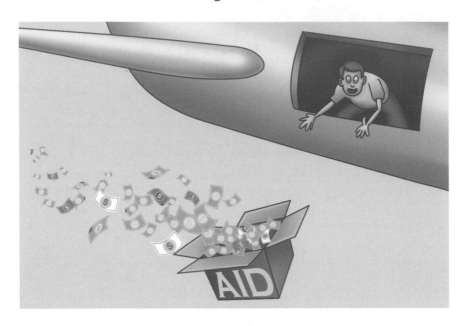

e) Using **Figure 10** and your own knowledge, describe the problems associated with giving aid. [2]

f) For a named lower income (developing) country, explain how globalisation and changing international trade have affected the process of development. [8]

Total marks for Question 6 = 21

END OF QUESTIONS

Collins

GCSE
GEOGRAPHY

Paper 3 Geographical Fieldwork, Issues and Decision-making

Time allowed: 1 hour 15 minutes

Materials

For this paper you must have:

- a clean copy of the pre-release resources booklet (except for **Edexcel A** and **Eduqas B**)
- a pencil
- a ruler.

Instructions

- Use black ink or a black ball-point pen.
- Answer all questions.
- Cross through any work you do not want to be marked.

Information

- The marks for questions are shown in brackets **[ ]**.
- The total number of marks available for this paper is 68.
- Spelling, punctuation and grammar will be assessed in Question **2A c)** or **2B d)** depending on your board and specification.

Make sure you have studied the resources booklet before completing this practice paper (except for Edexcel A and Eduqas B).

Name:

Question 1

Question **1A** is for **AQA**, **Edexcel B**, **Eduqas A**, **OCR A** and **OCR B**

Question **1B** is for **Edexcel A** and **Eduqas B**

Question 1A Issue evaluation

Study **Figure 1**, a world map showing the distribution of tropical rainforests.

Figure 1

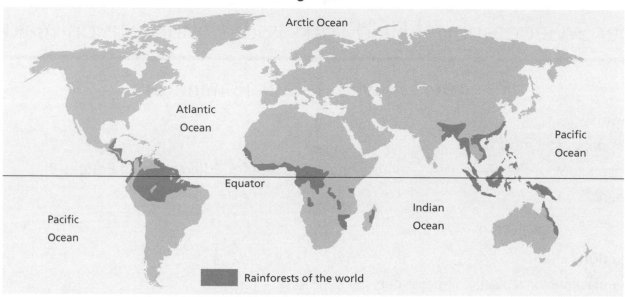

a) Using **Figure 1**, which **two** of the following statements are true?

Tick **two** options only. [2]

A There are significantly more rainforests in the Southern Hemisphere than in the Northern Hemisphere.

B There is a band of rainforests north and south of the Equator.

C The largest area of rainforest is in Australia.

D There are no rainforests in south-east Asia.

E The largest area of rainforest is in South America.

Study **Figure 2**, a table showing the 10 countries that have experienced the greatest loss of rainforest (in hectares) in the 21st century so far.

Figure 2

Country	Rank	Average loss 2010–2014	Trend
Brazil	1	2 347 727	Down
Indonesia	2	1 543 623	Up
DR Congo	3	778 348	Up
Malaysia	4	469 511	Up
Paraguay	5	406 785	Up
Bolivia	6	291 167	Down
Myanmar	7	207 677	Up
Madagascar	8	203 165	Up
Cambodia	9	187 893	Up
Peru	10	187 196	Up

b) Describe **two** links between the locations of the countries and the patterns shown in the chart. [2]

Study **Figure 3**, 'Threats to Rainforests and Biodiversity and their Impacts', in the resources booklet.

c) Explain why rainforests have such high levels of biodiversity. [6]

Question 1A continues on the next page

Study **Figure 4**, a graph showing mean global temperature changes since 1880.

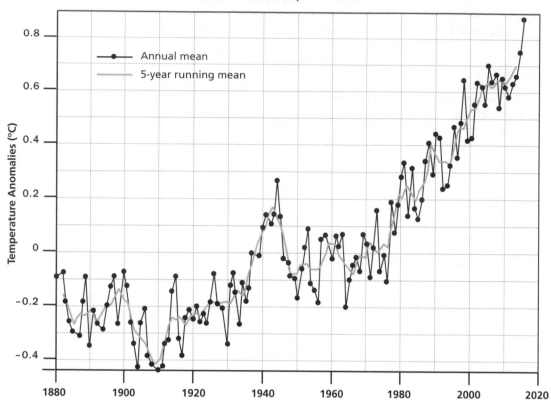

Figure 4

Global Land–Ocean Temperature Index

d) Describe the trend shown in **Figure 4**.   [2]

_____

_____

e) What was the difference in mean temperature between 1910 and 2010?   [1]

_____

f) In which decade was mean temperature change greatest?   [1]

_____

Total marks for Question 1A = 14

Question 1B Issue evaluation

This question is about UK national parks and potential conflicts within them.

Study **Figure 5**, a map showing national parks in the UK.

Figure 5

a)   Using **Figure 5**, state the number of national parks in England and Wales.   **[1]**

b)   Describe the distribution of English national parks.   **[2]**

Question 1B continues on the next page

Study **Figure 6**, a table showing information about the national parks.

Figure 6

National Park	Resident Population	Visitors per Year (million)	Visitor Spending per Year (£ million)
Brecon Beacons	32 000	4.15	197
Broads	6271	8.00	568
Cairngorms	17 000	1.50	185
Dartmoor	34 000	2.40	111
Exmoor	10 600	1.40	85
Lake District	41 100	16.40	1146
Loch Lomond & Trossachs	15 600	4.00	190
New Forest	34 922	Not available	123
Northumberland	2200	1.50	190
North York Moors	23 380	7.00	538
Peak District	37 903	8.75	541
Pembrokeshire Coast	22 800	4.20	498
Snowdonia	25 482	4.27	396
South Downs	120 000	Not available	333
Yorkshire Dales	23 637	9.50	406

c) Using **Figure 6**, state the national park that has the most visitors per year. [1]

d) Suggest **one** reason why some national parks receive many more visitors than others. [2]

e) Using the data in **Figure 6**, suggest why residents of the Broads National Park might be differently affected by visitors than residents of Exmoor National Park. [2]

Study **Figure 7**, a chart showing the relationship between the number of visitors per year and visitor spending at different national parks.

New Forest and South Downs are not shown as the information for those parks is incomplete.

Figure 7

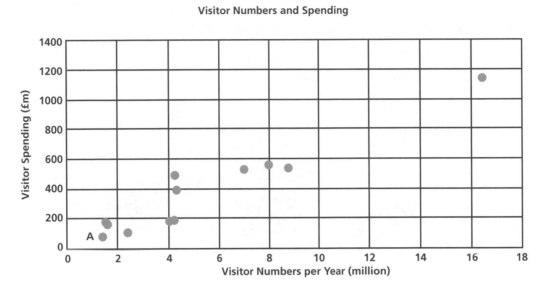

Visitor Numbers and Spending

f) Using **Figure 6** and **Figure 7**, identify which national park is shown by point A. [1]

g) Plot the point for the Yorkshire Dales National Park on **Figure 7**. [2]

h) Describe and suggest reasons for the pattern shown in **Figure 7**. [3]

Total marks for Question 1B = 14

Question 2

Question 2A is for AQA, Edexcel B, Eduqas A and OCR B
Question 2B is for Edexcel A, Eduqas B and OCR A

Question 2A Issue Evaluation

Study **Figure 8**, a map showing how the economies of countries will be affected by climate change, and **Figure 9**, a table showing GDP per capita in 2014.

Figure 8

Estimated Economic Impact of Climate Change by 2100

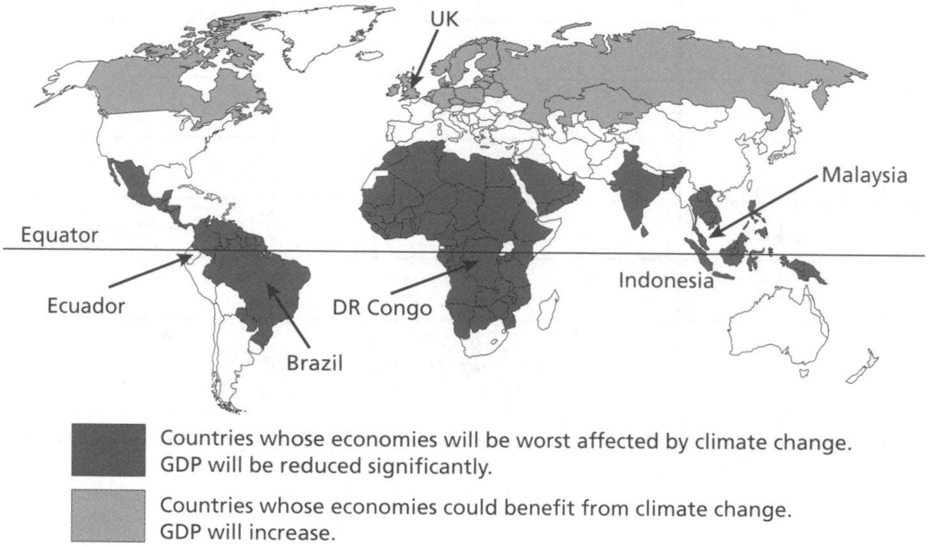

Countries whose economies will be worst affected by climate change. GDP will be reduced significantly.

Countries whose economies could benefit from climate change. GDP will increase.

Figure 9

Country	GDP per capita 2014 ($US)
Brazil	11 124
DR Congo	311
Ecuador	4657
Indonesia	3125
Malaysia	9069
UK	38 292

a) Rainforest destruction has been directly linked to climate change.

Describe the impact of rainforest destruction on the economies of different countries using the information given in **Figures 8** and **9**. [2]

b) Study **Figure 10**, 'The Challenges of Rainforest Conservation', in the resources booklet.

Explain possible conflicts that could arise over rainforests and their exploitation between the richer countries of the world and the countries where rainforests are found. **[6]**

....

....

....

....

....

....

....

There are many organisations which work to conserve the rainforests. They have developed a range of different conservation strategies to benefit the whole world, which also take into account the needs of local people and economic issues in the countries where the rainforests are found.

Study **Figure 11** and **Figure 12**, statements by two such organisations that need to raise awareness and money.

Figure 11

Organisation A

We are an NGO that works with local NGO conservation groups rather than governments. With donations from people around the world, we buy rainforest land to create reserves, which are passed on to the local groups.

We give scientific or technical advice and employ 'Rangers' who have a variety of roles: monitoring their area; keeping trails clear; maintaining fences; carrying out repair work after fires or storms; growing and planting trees; giving education talks; leading tourist walks; and assisting researchers. The presence of these Rangers deters poaching and other illegal activity and they earn a wage, which gives them independence and status within their community.

Our projects are small-scale but sustainable. We aim to protect the world's most biologically important and threatened habitats acre by acre.

Since being founded in 1989, we have funded partner organisations around the world to create reserves and give permanent protection to habitats and wildlife.

Question 2A continues on the next page

Figure 12

Organisation B

We are an NGO involved in conservation work. We work with governments, providing expertise and knowledge to help maintain protected areas that they have created. A typical scheme is working with individual native tribes in their own rainforest territory.

Donations from around the world provide boats, radios, fuel, border control training and aerial surveys. Surveillance has stopped illegal mining and mapped locations vulnerable to logging and fishing.

We are setting up non-timber, sustainable businesses with nuts, fruit, honey and essential oils. These provide jobs in harvesting, processing and transport.

We are dedicated to managing the things that we can control. We want societies to responsibly and sustainably care for nature and our global biodiversity, for the well-being of all humanity.

In this question, 4 of the marks awarded will be for spelling, punctuation and grammar and for your use of specialist terminology.

c) Which of the two organisations do you think will most effectively address the issues arising from threats to the rainforest?

Use evidence from the resources booklet and your own understanding to explain why you have reached the decision. **[12]**

Total marks for Question 2A = 20

Question 2B Issue evaluation

The national parks were set up in the 20th century. Two of their aims are to:
* conserve and enhance the natural beauty, wildlife and cultural heritage
* promote opportunities for the understanding and enjoyment of the special qualities of national parks by the public.

a) Suggest why some people might say that it is not possible to achieve both aims.   [2]

Study **Figure 13**, a photograph taken from a footpath near Ullswater in the Lake District National Park.

Figure 13

b) Using **Figure 13**, describe the attractions for visitors from an urban area.   [3]

Question 2B continues on the next page

Study **Figure 14**, a table showing some of the benefits and problems that having many visitors brings to the national parks.

Figure 14

Benefits	Problems
Jobs for local people	Damage to the landscape: litter, erosion, fires, disturbance to livestock, vandalism
Income for the local economy	Traffic congestion and pollution
Helps preserve rural services like buses, village shops and post offices	Local goods can become expensive because tourists will pay more
Increased demand for local food and crafts	Shops stock products for tourists and not everyday goods needed by locals
Tourists mainly come to see the scenery and wildlife, so there is pressure to conserve habitats and wildlife	Demand for development of more shops and hotels
	Jobs are mainly seasonal and low paid with long hours
	Demand for holiday homes makes housing too expensive for local people

National parks are very popular locations for day and residential school fieldwork activities.

Study **Figure 15** and **Figure 16**, photographs taken during a school field trip to the Yorkshire Dales National Park.

Figure 15   **Figure 16**

c) Describe the evidence for visitor pressure shown in **Figure 15**.   **[2]**

In this question, 4 of the marks awarded will be for spelling, punctuation and grammar and for your use of specialist terminology.

d) Using **Figure 14**, **Figure 16** and your own knowledge, discuss the contribution of fieldwork activities and trips to both the benefits and problems of visitors to national parks.　　[12]

e) There is an ongoing project to improve 'Brockhole', the Lake District National Park visitor centre at Windermere. One objective is to "…act as an example of sustainability".

Give **one** way in which this aim might be achieved.　　[1]

Total marks for Question 2B = 20

Question 3

Question 3 is for **all** boards and specifications

Question 3 Fieldwork

a) In the context of fieldwork, what do you understand by the term 'sampling'? [1]

b) What is a line sample? [1]

c) A school group is carrying out a questionnaire survey in a town centre about shopping. Explain which type of sampling would be most appropriate and reliable. [2]

Study **Figure 17**, a pie chart showing the results of a questionnaire survey about the type of housing in which respondents live.

Figure 17

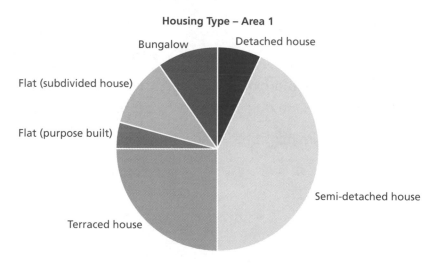

Housing Type – Area 1

d) Estimate the percentage of respondents from Area 1 living in terraced housing. [1]

Study **Figure 18**, a table showing the results of a similar survey carried out in a different part of the same town, and **Figure 19**, a divided bar chart produced by a student to show the results.

Figure 18

Housing type – Area 2	%
Detached house	1
Semi-detached house	23
Terraced house	46
Flat (purpose built)	10
Flat (subdivided house)	20
Bungalow	0

Figure 19

% Housing Type – Area 2

Detached house   Semi-detached house   Terraced house   Bungalow

e) Complete **Figure 19** to show the percentage of respondents living in the **two** different types of flat. [2]

f) What is the modal class of housing type in Area 2? [1]

Question 3 continues on the next page

Study **Figure 20**, a diagram showing visits to nearby towns by people living in villages to shop for clothes.

Figure 20

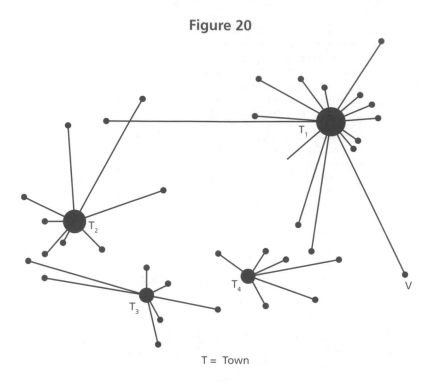

T = Town

g) Name the type of diagram shown in **Figure 20**. [1]

h) Suggest **one** reason why people from village V might travel to town T_1. [1]

i) A group of students is measuring bedload on a moorland river.

They are interested in how the long axis of bedload changes downstream, so they take 25 samples across the channel width at 10 locations starting from near the source. Their results are shown in the table.

Location	Long Axis of Longest Sample (mm)	Long Axis of Shortest Sample (mm)	Mean Length of Long Axis of all 25 Samples (mm)
1	360	94	250
2	275	78	176
3	343	81	201
4	212	55	153
5	246	37	131
6	197	32	53
7	148	28	51
8	112	14	29
9	73	11	17
10	44	5	8

The students want to present their results to show all three pieces of data at each location.

Draw a diagram to show how the information might be presented. You should not plot the figures. [3]

Study **Figure 21**, a map showing travel times by road (in minutes) to a large town from a number of locations in the surrounding area.

The 5, 10 and 25-minute isochrones (isolines joining up places of equal time) have been inserted.

Figure 21

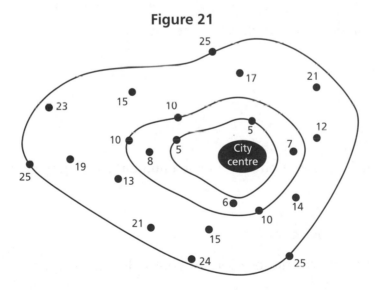

j) **Figure 21** shows that it is much quicker to travel into the town from some directions than others.

From which direction are the slowest routes?

Tick **one** option only. [1]

A   north east ☐   B   south east ☐

C   south west ☐   D   north west ☐

Total marks for Question 3 = 14

Question 4

Question **4A** is for **AQA, Edexcel A, Eduqas A** and **Eduqas B**

Question **4B (Coastal Fieldwork)** or **4C (River Fieldwork)** are for **Edexcel B, OCR A** and **OCR B** – answer **one** question only

Question 4A Fieldwork and decision-making

State the title of your fieldwork enquiry in which **physical** geography data was collected.

a) Describe **one** technique used in your enquiry to collect primary data. [2]

b) Name **one** presentation method used in your enquiry and justify its use. [3]

c) Justify a statistical method used in your enquiry. [3]

State the title of your fieldwork enquiry in which **human** geography data was collected.

d) State the location of your enquiry.

Explain the advantages of this location for your fieldwork enquiry. [3]

e) Describe **one** possible risk in carrying out your enquiry. [1]

f) Explain how this risk was reduced. [2]

....

....

g) What conclusions did your results enable you to make? To what extent were these in line with the geographical concepts or theories underpinning the enquiry? [6]

....

....

....

....

....

....

Total marks for Question 4A = 20

Practice Exam Paper 3

Answer **either** Question **4B** (Coastal Fieldwork) **or** Question **4C** (River Fieldwork)

Question 4B Fieldwork and decision-making (Coastal Fieldwork)

A group of students conducted an experiment to measure longshore drift.

They took 20 pebbles in a range of sizes and painted each size a different colour.

The pebbles were then put down in the swash / backwash zone, away from groynes, and a pole was stuck in to mark their position.

After a count of 50 waves, the students looked for the pebbles and measured the distances that the pebbles had travelled.

a)   Explain why pebbles in a range of sizes were used.   **[2]**

_____

_____

b)   The students made sure the pebbles were placed away from the groynes.

Explain what difference groynes would have made.   **[2]**

_____

_____

The first time this experiment was tried, 70% of the pebbles were found.

c)   Suggest reasons for only 70% of the pebbles being found.   **[2]**

_____

_____

Study **Figure 22**, a table showing some of the results from the experiment.

Figure 22

Pebble Size	Number of Pebbles Found	Minimum Distance Travelled (m)	Maximum Distance Travelled (m)
1 (smallest)	16	4	27
2	12	2	23
3	15	2	17
4	14	1	12
5 (largest)	13	2	15

d) Draw an annotated diagram/sketch to show how the information about distance travelled in **Figure 22** could be suitably presented. **[4]**

e) The students decided to use an 'average' distance travelled for each of the sizes of pebble.

Discuss the use of the mean, median and mode for their data. **[6]**

f) Give **four** precautions the students would have taken to ensure their safety when carrying out their experiment. **[4]**

Total marks for Question 4B = 20

Question 4C Fieldwork and decision-making (River Fieldwork)

A class of students is carrying out fieldwork to investigate downstream changes in size and shape of bedload.

They are working in five small groups. Each group carries out measurements at five different sites on the same river.

The methods of data collection for each group must all be the same.

a) State why it is important to decide on methods before taking any measures. **[1]**

The whole class stays at the first site to check that their agreed methods work before moving off to their own sites. They are careful to work facing upstream.

b) Explain why they should always face upstream. **[2]**

There is a discussion about how to decide which pieces of bedload will be measured.

Some students want to measure the bedload every 20 cm across the width of the river. Others want to take measurements at the same number of points across the channel at each site.

c) Explain the strengths and weaknesses of each method. **[4]**

Three measurements of each bedload sample are to be taken: longest, shortest and third axis.

d) Explain some of the problems that might be experienced when measuring bedload. **[6]**

Study **Figure 23**, the Power's Scale used by the students to look at change of shape.

Figure 23

Class 1	2	3	4	5	6
Very angular	Angular	Sub-angular	Sub-rounded	Rounded	Well-rounded

e) How is the data collected using the Power's Scale different from the data collected on bedload size?   [1]

Study **Figure 24**, a table showing the data for the first and last sites of the 25 sites surveyed.

Figure 24

Site		Biggest Sample (mm)	Smallest Sample (mm)	Median Sample (mm)
	Longest axis	760	8	327
1	Shortest axis	521	5	256
	Third axis	476	3	174
	Longest axis	20	2	5
25	Shortest axis	4	1	3
	Third axis	3	1	2

Question 4C continues on the next page

f) The students discuss how to present their data and decide to start with information about the median sample.

Using an annotated sketch/diagram, show a suitable technique for presenting the median sample data. **[4]**

g) There is an overnight storm forecast for the area of the river study. The students think it will provide an opportunity to see if increased rainfall affects the bedload.

Suggest an experiment the students might set up to test this idea. **[2]**

Total marks for Question 4C = 20

END OF QUESTIONS

Collins

GCSE
GEOGRAPHY

Resources for Paper 3 Geographical Fieldwork, Issues and Decision-making

Study the resources in this booklet before completing Practice Paper 3.

The resources for Paper 3 of your actual exam will be issued to you 12 weeks before the date of the exam.

This booklet contains two resources as follows:

- Figure 3 – Threats to Rainforests and Biodiversity and their Impacts: pages 240–244
- Figure 10 – The Challenges of Rainforest Conservation: pages 245–246

Figure 3
Threats to Rainforests and Biodiversity and their Impacts

Distribution of Tropical Rainforests

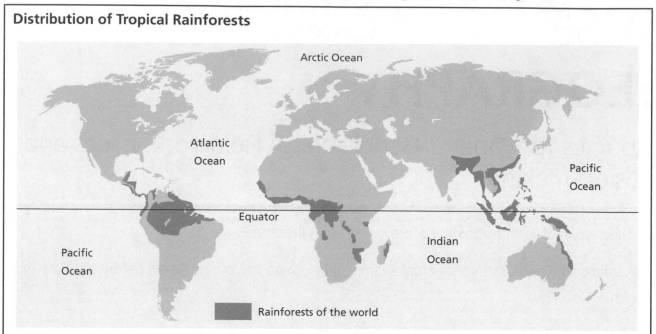

Biodiversity

Biodiversity generally refers to the range of species of plants and animals in a region or country. Countries containing tropical rainforest score more highly for biodiversity than those without. Indonesia is the most biodiverse country in the world.

Biodiversity Scores by Country

Country	Biodiversity score
Indonesia	1.00
Colombia	0.94
Mexico	0.93
Brazil	0.88
Ecuador	0.88
Iceland	0.10

Factors Affecting Biodiversity

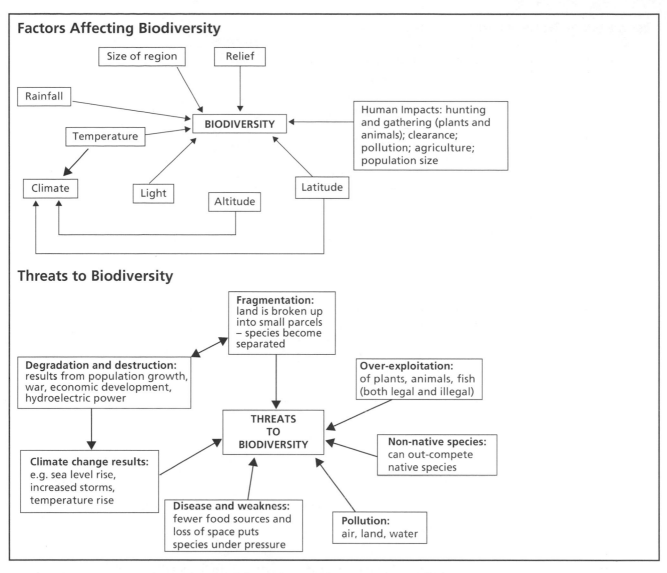

Threats to Biodiversity

Fragmentation of Forest as Logging Takes Place

Estimates of Tropical Forest Loss (hectares)

Country	Rank	Average loss 2010–2014	Trend
Brazil	1	2 347 727	Down
Indonesia	2	1 543 623	Up
DR Congo	3	778 348	Up
Malaysia	4	469 511	Up
Paraguay	5	406 785	Up
Bolivia	6	291 167	Down
Myanmar	7	207 677	Up
Madagascar	8	203 165	Up
Cambodia	9	187 893	Up
Peru	10	187 196	Up

Empty Forests

More than half of all tropical protected areas may be 'empty forests' — containing trees but few animals as a result of over-exploitation and uncontrolled hunting. As a result, animal species are in danger of extinction, tree species lose important seed dispersal and local people lose an important supply of protein.

Tree Diversity on Rainforest Alliance Certified Coffee Farms in El Salvador

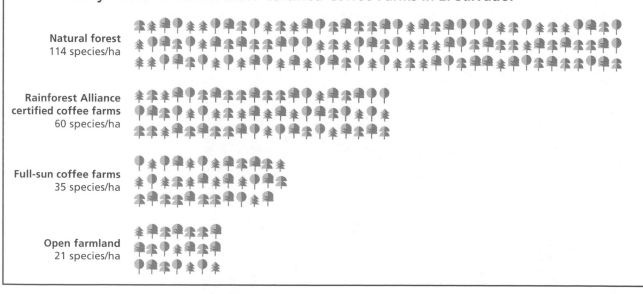

Natural forest
114 species/ha

Rainforest Alliance
certified coffee farms
60 species/ha

Full-sun coffee farms
35 species/ha

Open farmland
21 species/ha

Changes in Global Mean Temperatures

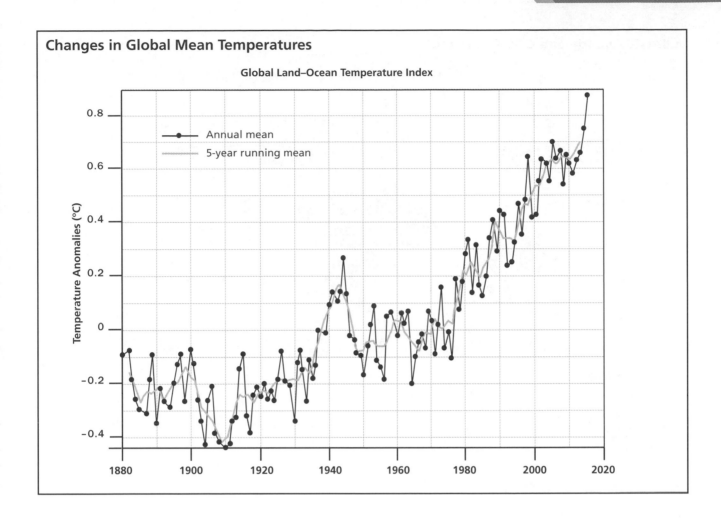

Climate Change and GDP Prediction

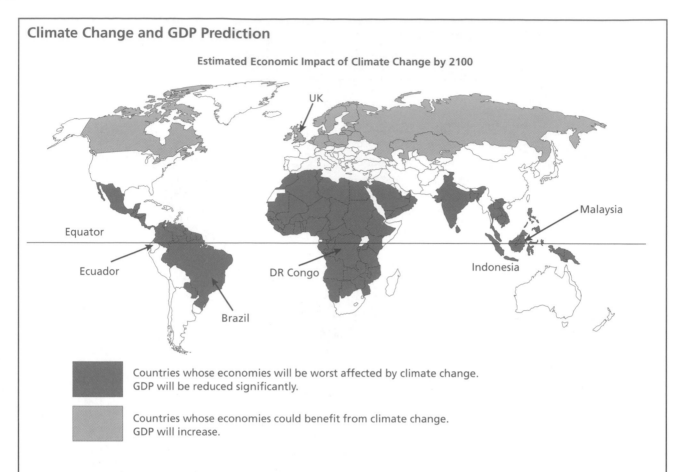

Estimated Economic Impact of Climate Change by 2100

Countries whose economies will be worst affected by climate change. GDP will be reduced significantly.

Countries whose economies could benefit from climate change. GDP will increase.

GDP per Capita in 2014 ($US) by Country

Country	GDP per capita 2014 ($US)
Brazil	11 124
DR Congo	311
Ecuador	4657
Indonesia	3125
Malaysia	9069
UK	38 292

Figure 10
The Challenges of Rainforest Conservation

The Value of the Forests

Tropical forests provide a range of goods and invaluable services. They are very important to the hydrological cycle (i.e. rain and water systems), and they maintain some of the world's most vulnerable soils. They are also one of the world's main carbon reservoirs. By absorbing carbon dioxide from the air, storing the carbon and providing oxygen, tropical forests act as the world's thermostat, regulating temperatures and weather patterns. The loss of tropical forests contributes to between 12% and 15% of all greenhouse gas emissions each year – about the same percentage as all of the world's trains, aircraft and cars combined. In some tropical countries, such as Brazil and Indonesia, emissions from deforestation can be as high as 60%, which is higher than emissions from all other sources. Replanting and rehabilitating secondary forests around the world has tremendous potential for offsetting greenhouse gas emissions. Furthermore, rehabilitated forest lands can attract eco-tourists and offer a habitat for native forest wildlife.

Rapid Tree Growth in a Cleared Area of Rainforest

In order to gain maximum access to sunlight, nutrients and water, new trees grow quickly. This means that young plants take a much greater amount of carbon dioxide from the atmosphere for photosynthesis, the process by which energy from sunlight is used to produce the sugars that the plants need to grow. In the best conditions, new-growth vegetation could take up to 11 times more carbon dioxide as old-growth forests. However, the long established old-growth rainforests have locked away a vast quantity of carbon over the decades and centuries.

Endangered Species

Jaguars need large areas of habitat in order to survive. In Brazil, they may have home ranges as large as 142 square kilometres.

They are formidable hunters. In fact, the jaguar's common name means 'the beast that kills its prey with one bound' in Indian traditions. Prey, which is both large and small animals, including tapir, peccary, birds and fish, will be stalked or ambushed, then dragged to cover. They will even eat snakes, turtles, porcupines and caiman.

Females raise two cubs that stay with their mother for about two years. Although jaguars are primarily thought to be nocturnal, they are actually crepuscular (mostly active around dawn and dusk), but can also be active during the day.

The rainforests of the Amazon Basin have the largest numbers of jaguars. Jaguars like water and are very good swimmers and are often found near rivers. They are also good climbers. In captivity, jaguars can live over 30 years, but in the wild they are unlikely to even reach half this age.

It has been estimated that jaguars are now only found in 46% of their former range. In the 1960s and 70s, 18 000 jaguars were killed per year for their fur, which caused a devastating decline in their population. As farms and ranches encroach further and further into wild jaguar habitat, these big cats will sometimes prey on domestic farm animals so are often shot by farmers and ranchers.

Habitat destruction means that jaguar populations are becoming increasingly isolated and are, therefore, more vulnerable. There is no legal protection of jaguars in Ecuador and the hunting of 'problem animals' is still allowed in Brazil, Guatemala, Mexico and Costa Rica.

Malaysia and Indonesia (the island of Borneo) were covered in tropical rainforest with a wide biodiversity. Much of the lower-lying area has been cleared for palm oil plantations. Palm oil is used in cosmetics and many foodstuffs and is one of their major exports. The two countries produce about 85% of world supplies. The trade brings in a lot of foreign exchange and provides thousands of jobs. Plantations need the forest to be completely cleared. Trees of commercial worth are taken and the rest are burned. The survival of the orang-utan is seriously threatened as they are native to the lowland forest. Fragmentation of the forest means they do not have enough territory. Palm oil production is predicted to massively increase by 2020.

Conservation

There are many organisations which work to conserve the rainforests. They have developed a range of different conservation strategies to benefit the whole world, which also take into account the needs of local people and economic issues in the countries where the rainforests are found.

Organisation A

We are an NGO that works with local NGO conservation groups rather than governments. With donations from people around the world, we buy rainforest land to create reserves, which are passed on to the local groups.

We give scientific or technical advice and employ 'Rangers' who have a variety of roles: monitoring their area; keeping trails clear; maintaining fences; carrying out repair work after fires or storms; growing and planting trees; giving education talks; leading tourist walks; and assisting researchers. The presence of these Rangers deters poaching and other illegal activity and they earn a wage, which gives them independence and status within their community.

Our projects are small-scale but sustainable. We aim to protect the world's most biologically important and threatened habitats acre by acre.

Since being founded in 1989, we have funded partner organisations around the world to create reserves and give permanent protection to habitats and wildlife.

Organisation B

We are an NGO involved in conservation work. We work with governments, providing expertise and knowledge to help maintain protected areas that they have created. A typical scheme is working with individual native tribes in their own rainforest territory.

Donations from around the world provide boats, radios, fuel, border control training and aerial surveys. Surveillance has stopped illegal mining and mapped locations vulnerable to logging and fishing.

We are setting up non-timber, sustainable businesses with nuts, fruit, honey and essential oils. These provide jobs in harvesting, processing and transport.

We are dedicated to managing the things that we can control. We want societies to responsibly and sustainably care for nature and our global biodiversity, for the well-being of all humanity.

Answers

TOPIC-BASED QUESTIONS

Pages 164–167: Natural Hazards

1. Pacific Ring of Fire [1]
2. **Any suitable answer, e.g.** South American plate and African plate; Eurasian plate and North American plate [2]
3. **Any suitable answer, e.g.** Pacific plate and North American plate; Nazca plate and South American plate [2]
4. **Any suitable answer, e.g.** Tohoku, Japan (2011); Haiti (2010); New Zealand (2011); Sichuan (2008); Northridge, USA (1994); Nepal (2015) [2]
5. The focus is the point underground where the earthquake originates [1], while the epicentre is the point on the surface directly above the focus. [1]
6. **Any suitable answer with reference to:** Plate boundaries; fault lines; measurement of fault movements; use of satellite imagery; frequency of previous earthquakes; tiltmeters; measuring gas release; unusual animal behaviour; difficulty of accurate prediction of timings [6]
7. **Any suitable answers, e.g.** Water; tinned food; wind-up radio; wind-up torch; batteries; first-aid kit; survival blanket; clothing; solar light. **[Answer should describe how three items can be helpful. Up to 6 marks]**
8. **Any suitable answer with a named example and impacts, e.g.** Tohoku, Japan (2011) or Asian Boxing Day tsunami (2004). [1] Impacts: large expanse of coast affected; worse closer to epicentre of earthquake; fatalities; loss of settlements and roads; extreme flooding; damage to ships; effect on nuclear reactor at Fukushima (2011); effect on water supplies. **[Up to 5 marks]**
9. a) True [1]
   b) True [1]
   c) True [1]
   d) True [1]
   e) False [1]
10. Teleconnections [1]
11. a) **Any suitable answer making the key point that the storms are not inland or along the Equator, but mainly over ocean water between 5° and 20° north or south.**
    Pacific Ocean: east Pacific but mainly in the west Pacific near the Philippines and north-east Australia; north Atlantic; Indian Ocean (north and south of the Equator); rare in south Atlantic [4]
    b) Areas with higher sea surface temperatures tend to have the most tropical storms [1] but this is not the case along the Equator [1]. They do not form over land [1] and rapidly lose intensity over land because they lose their supply of warm, moist air. [1]
12. It is in the middle of the Pacific Ocean, a very long way from large urban or industrial areas [1]. This means that there is no bias in the measurements. [1]

Pages 168–169: Ecosystems

1. **Any suitable answer, e.g.** Deforestation in tropical rainforests [1], as the removal of trees destroys the habitats of countless organisms [1], thus leading to potential imbalance [1].
2. **Any two from:** Northern Atlantic Ocean; northern Pacific Ocean; Australia; off the coast of southern South America; southern Africa; south-western Indian Ocean [2]
3. **Any suitable answer, e.g.** Otter [1]
4. For stability [1]; to absorb nutrients from the leaf litter [1]
5. Salinisation [1]
6. By encouraging migration of non-native species [1]
7. Extreme tourism [1]
8. **Any three from:** Heathland; moorland; wetland; woodland [3]
9. Stratification [1]
10. **Any four developed points, e.g.** Road building leads to increased settlement; trees are cut down to make way for farms; trees are removed for access to mineral deposits; hydroelectric power schemes mean large-scale removal of trees. [4]
11. **Any three developed points, e.g.** Along and close to the Tropics of Cancer and Capricorn, such as the Sahara; between 15 and 30 degrees north and south of the Equator, such as the Namib; on the south-western edges of continents; on the leeward side of some mountain ranges, such as the Gobi. [3]
12. **Any named species and specific adaptation, e.g.** The saguaro cactus has no leaves so it does not lose water through transpiration; fennec foxes have large ears to radiate heat; quiver trees have fleshy stems to conserve water. [2]

Pages 170–173: Physical Landscapes in the UK

1. **Any four from:** Granite is formed deep underground [1]. Pressure from overlying material was greater where the joints [1] of granite are closer together. Removal over time of the surface [1] revealed the more resistant areas [1], which have rounded stones [1] on top where the corners of granite blocks have been weathered more than the faces [1].
2. Glacier A [1] because it is thicker and so puts more pressure on to the bedrock [1]. It also contains more debris which can be used for abrasion [1].
3. **Any suitable answer which includes the first point here and any two others:** Ice in the uplands is moving / has energy [1] so is eroding [1]. On lowland areas, ice has lost energy [1] so deposits material [1] but can 're-advance' [1] and so erode the landscape [1] and the depositional landforms [1]. **[Up to 3 marks]**
4. Groynes are built at right angles / perpendicular to the beach [1] so stop sand/material as it is carried along the shore [1].
5. a) Line A: summer   Line B: winter   **[1 for both]**
   b) **Any suitable answer with reference to wave type and action and result, e.g.** In winter there are stronger / more powerful waves / plunging waves / frequently destructive waves [1] so beach material is removed [1] and the profile is lower. Plunging waves drag material down the beach/ offshore [1], creating the ridge [1]. **[Up to 3 marks]**
6. **Any suitable answer, e.g.** Both types push material up the beach [1] but surging waves are more often on steeper beaches [1] and create a ridge of material [1], whereas spilling waves are generally on gentle [1] beaches and push material onshore [1], creating a higher beach [1]. **[Up to 2 marks]**
7. Rocks / Rock structure is parallel to the coastline. [1]
8. **Any suitable answer which refers to processes affecting both headlands and bays, e.g.** Headlands get worn back [1] by erosion / hydraulic action and abrasion [1] so stick out less [1], whereas bays have material deposited in them / have beaches in them [1] that build out the coastline [1]. **[Up to 4 marks]**
9. a) The cliff collapses [1] and is not able to maintain its height and/or shape [1].
   b) Rocks will absorb water / become saturated [1], so they are too heavy to be supported [1].
   c) Stormy weather means there will be stronger waves / more plunging waves [1], which attack the cliff foot [1], causing collapse.
10. **Any suitable answer which both describes the action [up to 2 marks] and the results [up to 2 marks]:**
    Action: abrasion is the scouring action of material carried in waves [1] against the cliff [1], especially the cliff foot [1].
    Results: this can create a notch / weaker area [1] which cannot support the cliff [1], so it can collapse [1]. Cliffs recede / wear back [1] and a wave cut platform emerges [1].
11. a) **Any suitable answer giving a named area and mentioning the physical reason for the protection and the human context, e.g.** Glacial till cliffs along the Yorkshire coast [1] are easily weathered and are eroding between 2 and 3 metres per year [1]. The B1242 road, connecting east coast settlements, was at risk [1].
    b) **Any suitable answer for the same example, e.g.** Two rock groynes of Norwegian granite [1] were constructed at Mappleton to trap sand [1] and steepen the beach [1]. **[Up to 2 marks]**

Answers

12. **Any suitable answer, e.g.**
    Location of place/area: Selwicks Bay is at Flamborough in East Yorkshire, facing the North Sea / north-east of Bridlington. Selwicks Bay is the most easterly bay on Flamborough Head **[2 marks for any two clear locational points]**

    Erosion feature: The stack at Selwicks Bay. It is at the southern end of the bay. It is about 20 m high and composed of chalk with just a little bit of glacial till at the top. **[2 marks for any two specific/clear descriptive points]**

    Deposition feature: The beach at Selwicks Bay. Selwicks Bay beach is backed by chalk cliffs. It is pale, fine sand with scattered boulders, cobbles and pebbles of mixed rock. At low tide, the wave cut platform is visible. **[2 marks for any two specific/clear descriptive points]**

13. **Any suitable answer with qualified ideas, e.g.** Loss of permeable surfaces / increased impermeable surfaces **[1]**, so less infiltration **[1]** and faster overland flow **[1]**. Pipes, drainage and sewerage **[1]** speed up water through the ground **[1]**. River channels may be built over / lost **[1]**. Construction changes surface material **[1]**, which could affect interception or throughflow **[1]**. Trees/vegetation likely to be lost **[1]** so less interception **[1]**. **[Up to 4 marks]**

14. It is a river level / water level gauge **[1]**, which suggests that this river often floods / rises much higher than it is on the photo **[1]** and that there are buildings/places at risk. The river could rise up to 6 feet higher than on the photo **[1]**.

Pages 174–177: The Human Environment

1. A city with a population in excess of 1 million people **[1]**
2. A city with a population in excess of 10 million people **[1]**
3. **Any three from:** Contraception; financial reasons; fewer infant deaths; women pursuing careers; family planning; mechanisation of agriculture **[3]**
4. A **[1]**; C **[1]**
5. **Any suitable answer with three points, e.g.** Escape traffic congestion and overcrowding; lower house prices; cleaner air; ability to work from home **[3]**
6. **Any suitable answers, e.g.** The creation of Queen Elizabeth Park; new housing; new sports facilities; the cleaning up of old industrial areas; improved transport systems **[Up to 4 marks]**
7. **Any suitable answer which describes four points, e.g.** Overcrowded living conditions; lack of clean water; no sewage disposal; few services; no land ownership **[4]**
8. **Any suitable answers, e.g.** Harlow; Cumbernauld; Welwyn Garden City; Skelmersdale **[3]**
9. **A:** CBD **[1]**
   B: Zone of transition **[1]**
   C: Residential (lower class) **[1]**
   D: Residential (middle class) **[1]**
   E: Residential (upper class) **[1]**
   F: Rural–urban fringe **[1]**
10. **Any suitable answer, e.g.** Renewable energy use for housing and industry; efficient recycling; energy generated from waste; development of integrated public transport system; creation of cycleways **[4]**
11. **Any suitable answer which describes how income can be generated from non-agricultural ventures, e.g.** Bed-and-breakfast facilities to offer accommodation to tourists; ice-cream manufacture to provide refreshments; shops selling farm produce; campsites to make further use of fields **[2]**
12. **Any suitable answers, e.g.** Improved healthcare; new medical treatments and drugs; fewer manual jobs; better diet **[3]**
13. **Any suitable answers, e.g.** Farming; forestry; mining; oil exploration **[2]**
14. **Any suitable answers, e.g.** Any IT roles; research and development; software development **[2]**
15. **Any suitable sketch:** The shape should be a clear triangle. Labels could include: gender, age groups, steep sides, wide base, concave shape, narrow top – and these should be linked to birth rate / death rate changes and life expectancy. **[6]**

16. **Any suitable answers, e.g.** Pollution; overfishing; aquaculture; coral bleaching; lack of protection legislation; shipping; oil and gas exploitation; rising sea levels and rising temperatures connected to climate change **[5]**

17. Long-term aid – Aid that solves a problem so that it does not occur again
    Voluntary aid – Aid that comes from charities
    Multi-lateral aid – Aid that is provided by a group of countries
    Short-term aid – Aid that solves an immediate crisis
    Tied aid – Aid that comes with conditions
    [4 marks if all correct; 3 marks if three correct; 2 marks if two correct; 1 mark for one correct]

Pages 178–179: Resources

1. A dam is constructed for two or more purposes **[1]**, e.g. storage / flood control / navigation / power generation / recreation **[1]**.
2. **Any four from:** In the 1960s, most of the UK's electricity was generated using coal, oil and a small amount of nuclear **[1]**. Gas was also made from coal **[1]**. Now around half of the UK's electricity is made from gas **[1]**, the rest from coal, nuclear and renewables **[1]**. More fuel-efficient cars mean that only the same amount of oil is used for transport despite there being three times as many cars **[1]**. **[Up to 4 marks]**
3. Not all countries have the conditions that are right for power generation from renewable resources **[1]**, such as strong sunlight **[1]**, windy conditions **[1]** or steep mountains (for HEP) **[1]**. **[Up to 3 marks]**
4. **Any suitable answers, e.g.** Farmers are stuck in the cycle of poverty **[1]**; lack of investment in agriculture due to lack of funds or government corruption **[1]**; climate change and drought **[1]**; civil war leads to displacement **[1]**; fluctuations in crop prices **[1]**; food wastage **[1]**; poor storage and distribution networks **[1]**; rapidly rising populations **[1]** **[Up to 4 marks]**
5. **Any three from:** Organic farming is a sustainable method of food production **[1]** that does not use synthetic pesticides or fertilisers **[1]**. It instead uses organic products (plant and animal waste) to fertilise **[1]** and biological solutions for pest and disease control **[1]**. **[Up to 3 marks]**
6. Water held underground in the soil or in spaces in rock **[1]**
7. Fuels produced directly or indirectly from organic material including plant materials and animal waste **[1]**
8. **Any suitable answers, e.g.** There are many different groups of countries: some, like those in the Middle East and North Africa are arid and have shortages of water **[1]**. Others like South Africa and India **[1]** have poor infrastructure and are not able to maintain supplies to their population **[1]**. Other countries like the UK and China **[1]** have large population densities **[1]** and water in some areas may be in short supply due to high demand **[1]**. **[Up to 4 marks]**
9. **Any suitable answer, e.g.** War against other countries (e.g. Iraq) **[1]** or civil war (e.g. Syria, Libya) **[1]**; political regimes might not want to trade with western countries; embargoes from the UN or EU **[1]**; gradual exhaustion of oil supplies (e.g. Bahrain) **[1]**. **[Up to 3 marks]**
10. Geothermal energy is found mainly in areas where there is volcanic activity **[1]** (such as in Iceland).
11. **Any suitable answer, e.g.** France **[1]**

Page 180: Geographical Applications

1. **Answers will vary. [Up to 9 marks + 3 for SPaG]**

Answers

PRACTICE EXAM PAPERS

Extended Response Questions

When your exam papers are marked, the total mark you are given for each extended response question (usually worth 4–12 marks depending on the exam board) is based on the overall quality of your answer, including:

- the level of knowledge and understanding of the subject / issue that you demonstrate
- the level of accuracy, including correct use of relevant specialist terms
- how well you use evidence/data/examples to support your ideas
- how well developed your answer is:
  - o Is it a balanced discussion / does it consider a range of viewpoints?
  - o Is the meaning clear and easy to understand?
  - o Is it structured in a logical way that is easy to follow with a clear conclusion?

Don't be misled into thinking that spelling, grammar and punctuation are only important in the questions where marks are specifically awarded for these elements. Good grammar is essential in all written answers to ensure that your meaning is clear and your answer is not misunderstood. A good answer will always consist of well-developed sentences that use connectives, e.g. *therefore*, *however*, *as a result* (rather than a basic list of ideas written as simple clauses / part sentences).

For the purpose of this book, to help you mark your own answers, we have identified the key points that should be addressed in each extended response answer. However, it is important to remember that you will only be awarded full marks in the exam if these points are communicated in a clear, accurate and well-developed way (as outlined above). An example marks table that illustrates this has been included for Paper 1 Question 1A f). A similar scale will be applied to all extended response questions in the exam.

Spelling, Punctuation and Grammar

The questions that will be used to specifically assess the accuracy of your spelling, punctuation and grammar will be clearly identified on your exam papers. For each of these questions, three or four marks are available as follows:

- **High performance: 4 marks**
  - o spelling and punctuation is consistently accurate
  - o well-written with excellent use of grammar, so meaning is always clear
  - o a wide range of specialist terms are used appropriately
- **Intermediate performance: 2–3 marks**
  - o spelling and punctuation is considerably accurate
  - o good use of grammar, so meaning is generally clear
  - o a good range of specialist terms are used appropriately
- **Threshold performance: 1 mark**
  - o spelling and punctuation is reasonably accurate
  - o reasonable use of grammar (overall, any errors do not significantly hinder understanding of the answer)
  - o a limited range of specialist terms are used appropriately
- **No marks awarded: 0 marks**
  - o no answer has been given
  - o the answer does not relate to the question
  - o spelling, punctuation and grammar does not reach the threshold level (errors mean that the answer cannot be properly understood)

Paper 1: Pages 181–200

Question 1A

a) B [1]; C [1]

b) North American; Pacific [1]

c) **Any one from:** Magnitude refers to the strength/energy of an earthquake; is measured on the Richter scale [1]; **Plus,**

any one from: 7.8 is a strong/powerful earthquake; has the potential to cause a lot of damage [1]

d) Landslide / Landfall / Rock fall [1]

e) **Any two from:** Roads or other transport / communications links could be blocked; preventing evacuation / rescue / arrival of emergency services [2]

Farmland could be covered/destroyed/ altered; so there will be a shortage of food or a fall in farmers' earnings [2]

Houses could be covered / made structurally unsafe; leading to injuries and/or fatalities / creating homelessness [2]
[It is important to give the impact of the landslide and the effect on life for 2 marks in each case]

f) **Your answer must:**
- refer to the stimulus material and case studies, which should be named and located
- describe and explain both hazards and benefits
- include a balance of hazards and benefits
- demonstrate good use of spelling, punctuation and grammar.

Benefits to include:
- good soils, e.g. Mt. Etna has orange groves and specific varieties of apricot and blood orange are only found there; Mt. Rainier is famous for its cherry and apple orchards
- tourism, e.g. Mt. St. Helens
- scientific study
- geothermal energy, e.g. Iceland

Hazards to include:
- respiratory disease/illness, e.g. Mt. Pinatubo
- evacuation difficulties, e.g. Montserrat
- damage to farming – crops and livestock
- damage to transport infrastructure, e.g. roads, rail links, bridges
- pollution of water supplies
- damage to property

Example marks table (at each level, marks move up with increased detail):

1–3 marks	• Description of hazards or benefits only.
4–6 marks	• Both hazards and benefits are mentioned, although imbalanced. • Attempts to explain hazards/benefits as well as describe them. • Use of examples.
7–9 marks	• Both hazards and benefits are described and clearly explained (some imbalance is allowable). • Relevant examples given. • Good use of examples.

Question 1B

a) Synoptic [1]

b) 1030 mbs [1] **[Unit must be given for mark]**

c) The closer together isobars are, the stronger the wind [1]. On the chart shown, they are far apart, showing calm conditions [1].

d) Clear / Cloud free [1]

e) Sea breezes (keep temperatures down) [1]

f) **Your answer must:**
- refer to the stimulus material and case studies, which should be named and located
- describe and explain the negative impacts
- consider economic, social and environmental issues
- demonstrate good use of spelling, punctuation and grammar.

Answers

Negative impacts may include:
- health issues such as heatstroke, especially among the very young and elderly
- water shortages / rationing such as hosepipe bans
- greater demand for water to water gardens and irrigate crops
- crop failure leading to decreases in farmers incomes
- food shortages leading to increased food prices
- road surfaces melt and train lines buckle leading to transport disruptions
- wildfires breaking out.

Example marks table (at each level, marks move up with increased detail):

1–3 marks	• Description of negative impacts only.
4–6 marks	• Attempts to explain negative impacts as well as describe them. • Use of locations.
7–9 marks	• Negative impacts are described and clearly explained. • Relevant examples given. • Good use of locations.

g) Convectional rainfall **[1]**

h) Hot, dry conditions caused the ground to bake hard / made the surface impermeable / unable to absorb water **[1]**, so the rain flowed quickly over the surface **[1]**.

i) Fog / Very cold temperatures **[1]**

Question 2A

a) entirely **[1]**; south **[1]**; some **[1]**

b) Hot desert **[1]**

c) Three arrows must be added, pointing left to right, from green plant to caterpillar, caterpillar to bird and bird to cat **[1] [All three arrows are needed for the mark]**

d) Green plant **[1]**

e) An arrow drawn from 'rainfall intercepted by trees' to 'oxygen' (oxygen is an output of the cycle) **[1]**

f) Loss of water from a plant **[1]**

g) Much hotter and wetter conditions / faster reactions due to greater heat and moisture **[1]**

h) Weathering **[1]**

i) **Your answer must:**
- include both plant and animal adaptations (but balance not necessary)
- include at least two well-developed sentences
- make a clear link between each adaptation and the weather/climate, e.g. snakes that can swim to survive when rivers flood due to heavy rainfall; drip tips on leaves allow rapid shedding of rainwater, preventing leaf pores from getting clogged; thick, leathery leaves are able to withstand intense heat, which would cause thinner leaves to become too flaccid.

[A maximum of 3 marks can be awarded if only one element is covered or if the adaptations are described but not explained]

Question 2B

a) Decomposition / Breaking down of rock in situ **[1]**

b) Frost shattering / freeze-thaw action, in which water enters cracks in the rock and freezes and expands, putting pressure on the rock so it eventually breaks away. **[3] [The term 'frost shattering' / 'freeze-thaw action' does not need to be used, provided there is a full and accurate description]**

c) Vegetation has grown, showing that they are not changing/moving. **[1]**

d) **Any one from:** Lake District; Peak District; Snowdonia in Wales; north-west Scotland **[1]**

e) **Any one from:** Gentler slopes; complete vegetation / rough grassland coverage **[1]**

f) 1D, 2A, 3B, 4C **[3] [1 mark for one correct; 2 marks for two correct]**

g) **Your answer:**
- is likely to mention wetter upland areas / relief rainfall and wetter westerly areas / depressions
- might discuss drier areas.

(Good answers may look at a west/east split and include both types of rainfall or describe rainfall types and then locate them.)

Example: Higher land receives more rain / is wetter than lower **[1]**. Rain-bearing winds approach from the Atlantic **[1]**, which is to the west. Depressions form over the Atlantic **[1]** so affect the west more **[1]**. There is relief rain in the west because there is higher land / the Pennines / Lake District **[1]**, which causes air to rise and cool, and rain to form **[1]**. Eastern areas are in the rain shadow / lee of the higher land **[1]**. Air falls and warms, so is able to hold moisture **[1]**.

Question 3

a) **Your answer must:**
- name a large-scale ecosystem / biome
- give named examples of both plants and animals found in that ecosystem (fairly balanced)
- try to focus on species that are specific to that ecosystem
- make clear links between the plants and animals and the climate (specific temperature, amount of rainfall, etc.), i.e. detailed explanations of the adaptations that allow them to survive/thrive there.

Example, tundra biome: Arctic tundra regions are very cold, windy and dry with deep layers of permafrost; only the top 50 cm thaws in the summer. Average temperatures for the year are –6 to –12°C. They have 8 months with temperatures below zero and a high of only 10°C in summer. Winters are dark but there is 24 hours of daylight in summer. Precipitation is only 100–150 mm per year, most of which falls as snow in the summer. There are only 50–60 days warm enough for plant growth. Plants are low-growing because it is so windy. Mosses and lichens are typical plants. There are also grasses, sedges and dwarf willow. More sheltered areas have bilberry and crowberry. Flowering plants such as saxifrage and Iceland poppy must germinate, grow, flower and seed within six to seven weeks but they can photosynthesise efficiently in the 24-hour daylight. There are migrant animals, e.g. caribou and reindeer, which move into the tundra for summer grazing. The musk-ox survives because of a very thick coat. The tundra receives many migrating, wading and ground-nesting birds and geese which are hunted by arctic fox. Bog areas form in the summer, when just the top layers thaw out. Thousands of insects live on these, providing food for the birds.

b) **Your answer must:**
- name a large-scale ecosystem / biome
- give examples of specific threats to that ecosystem
- try to avoid generalised statements, e.g. too many tourists
- make clear links between the threats and management techniques, i.e. detailed explanations of the techniques being used and how they counteract a specific threat.

Example, savanna grasslands biome: The savanna is mainly grassland with trees. If left untouched, savanna grasslands house herds of herbivores, such as zebras, gazelles and elephants, which are the prey of carnivores such as lions and cheetahs. There are scavengers such as vultures and hyenas.

Fires are a natural occurrence but are also deliberately started by farmers to clear more land. Trees are also cut for fuel. Removing tees leads to loss of soil. There is illegal poaching of wild animals for food and export. With low human population densities, savanna grasslands stay healthy. Traditional nomadic herding was part of the ecosystem. Parks such as the Serengeti, Ngorongoro and Tarangire have been set up to protect animals. Wardens are employed to stop illegal poaching. The worldwide ban on trade in ivory was established to protect elephants and rhinoceros. Creating boreholes for water outside the protected areas was meant to preserve those areas but has led to a concentration of people and cattle and more cutting of trees. 'Green tourism' is encouraged. This provides jobs for locals, increases awareness of pressures on the savanna and provides income so that poaching is less attractive.

Question 4

a) Attrition; Waves crash bits of rock together and, as they are moved on a beach, they become smoother and rounder [2] **[The word 'attrition' does not need to be used, provided there is a full and accurate description]**

b) Slumping [1]

c) Clay / Boulder clay [1]

d) Sea arch [1] (or 'arch')

e) Headland: E [1]; Sandy beach: D [1]; Stack: A [1]

f) **Any two from:** Swash is stronger than backwash / backwash weaker than swash; more material is brought onto a beach than is removed; surging or spilling type; usually low height <1 metre; long wave length, up to 100 m between crests; long wave period, 6–8 per minute [2]

Question 5

a) Some of the load is large in size, so there must have been much more water at some point to transport the rocks downstream. [1]

b) There is moss/vegetation on some of the rocks in the channel, so they must have been exposed for a long time. [1]

c) A sudden change of gradient in the downstream course of a river [1]

d) **Your answer must:**
   • consist of four or five well-developed sentences
   • cover both formation and associated features
   • include examples of waterfalls and their particular situation (rock types, etc.)

   Waterfalls begin:
   • at faults / fault lines
   • on sides of glaciated valleys
   • as a result of tectonic uplift / a fall in base level / differential weathering.

   Associated features could include the following and specific erosion processes should be mentioned for each:
   • plunge pool
   • gorge.

e) Outer bend / At the outer edge [1]

Question 6

a) The average weather conditions [1]

b) The times of year when different seasons are expected to happen are changing / the UK is experiencing shorter / warmer winters [1]; plants are coming into growth and flowering earlier / butterflies are seen earlier in the year; earlier spring is allowing some species (e.g. hedgehogs) to have two litters of young [1] **[2 marks for any correct statement relating to how seasons are changing in the UK, plus an example]**

c) Overall a downward trend / steady though irregular fall [1]

to almost 15 metres below the start height of glaciers [1]. More rapid reduction/decline since 2000 [1]. **[Alternative answers must use figures from the chart to show understanding and give some detail of the trend]**

d) C [1]; D [1]

e) Using logged trees for fuel / Burning to clear land instantly releases carbon dioxide into the atmosphere [1]. Deforestation means fewer trees are absorbing the carbon dioxide being released [1]. **[Alternative answers gain 1 mark for referencing carbon dioxide and 1 mark for correctly linking it to the greenhouse effect]**

f) **Any two from:** More storms / intense rainfall; decreasing summer rainfall; increasing winter rainfall [2]

g) **Your answer:**
   • must state specific effects of climate change
   • must, for each effect, give a named location
   • must explain how the quality of life is being negatively affected for people living and working at that location
   • might mention the effects on transport, farming, homes and employment
   • should include two well-developed points or three clear, located ideas.

   Example: Stormier winters in the north and Scotland [1] could make transport difficult and arable farming crops more likely to fail [1]. Likely flooding of low-lying land near coasts, e.g. in East Anglia [1], means people may need to move house or have difficulty selling or insuring their property [1]. Storms will increase coastal erosion, e.g. along Holderness coast [1], leading to loss of homes and businesses and leaving some people homeless and/or unemployed [1].

h) **Accept any relevant single idea that is clearly described, e.g.** There are chalk cliffs in the south of England [1]; chalk is formed in tropical conditions, which shows that the climate must have been tropical in the distant past [1]. **[1 mark for stating evidence (e.g. vegetation/pollen analysis; documents / paintings of 'Little Ice Age'; rock types that form in particular conditions, etc.) and 1 mark for explaining how it shows that the UK climate was different previously]**

Paper 2: Pages 201–214

Question 1A

a) **Any two from:** Slow increase in urban population from 1960 to 2010; slow decline in rural population from 1960 to 2010; crossover in 2008, whereby there are more urban dwellers than rural [2]

b) **Any two from:** Rural to urban migration; growth of manufacturing industry; climate change; high birth rates [2]

c) Asia [1]

d) **Your answer:**
   • must include four or five well-developed sentences
   • must explain how each issue mentioned creates challenges
   • might reference infrastructure; transport; healthcare; education.

   Example: Investment in infrastructure, such as sanitation, is patchy [1]. As a result, waterborne diseases spread easily [1]. Transport links are poor, so businesses struggle to expand [1]. There is a lack of investment in healthcare and education [1], therefore death rates are higher [1], and local people are not equipped to build a better life for themselves [1].

Question 1B

a) **Any one from:** Literacy rates are generally higher in urban areas; the proportion of the population working in agriculture is higher in rural areas; the proportion of the population with access to clean drinking water is generally higher in urban areas **[Accept any other correct difference]** [1]

Answers

b) **Any two from:** Poor transport; hot climate; few job opportunities; poor sanitation; few schools; a traditional way of life [**Accept any other valid reason**] [2]

c) **Any two from:** An illegal settlement built by its residents; the housing is made out of makeshift materials; often found on the edge of cities; unplanned; high density of inhabitants [2]

d) **Your answer:**
   - must include four well-developed sentences
   - must name a specific rural area in the UK, e.g. east Devon
   - is likely to cover negative impacts of technological change
   - is likely to compare rural areas to urban areas
   - should clearly explain how service provision is affected in rural areas, using examples where appropriate, e.g. due to good access to the Internet, people are doing their shopping online and having it delivered, meaning they are not using local services as much and these are therefore closing.

Question 2A

a) A [1]

b) **Your answer must:**
   - name a regeneration project
   - include at least four well-developed sentences
   - mention both social and economic improvements.

   Example: The Liverpool One retail development in Liverpool's central business district [1] created a huge number of jobs, mostly in the retail and food service industries [1]. The project also helped to alleviate poverty in the city [1] as many of the jobs created were semi or unskilled and, therefore, open to all [1]. New transport links have helped businesses develop [1]. New hotels and visitor attractions have drawn tourists and therefore new wealth into the city [1].

Question 2B

a) Sustainable [1]

b) **Your answer must:**
   - name an out-of-town shopping centre
   - include at least four well-developed sentences
   - mention both costs (e.g. taking business away from town centre shops) and benefits (e.g. creating many jobs) to the local community.

   Example: The Trafford Centre on the edge of Manchester [1] created lots of jobs mainly in retail and food industries [1]. Many of these jobs were semi or unskilled jobs and therefore suitable for local residents [1]. Most shoppers travel using a car and this can cause congestion on local roads [1]. This creates noise and/or air pollution in the local area [1]. Many local people use the shops at the Trafford Centre rather than using their local shops, leading to a fall in trade in local shops. This has caused some local shops to close down, leading to unemployment [1].

Question 3

a) A [1]; B [1]

b) **Your answer:**
   - must include four or five highly-developed sentences
   - must name a city/location (1 mark will be lost if no named example is given)
   - must explain the strategies being used there to reduce traffic congestion
   - might comment on the effectiveness of the strategies
   - must use geographical vocabulary accurately
   - must demonstrate good use of spelling, punctuation and grammar.

   Example: The BRT [1] in Curitiba, Brazil [1], subsidises bus transport within the city [1], therefore dissuading lower income workers from using cars for transport [1]. Tubular bus stops, where passengers pay before they get on [1] mean the buses spend less time idling [1]. Bus routes are distributed across the city and cross at terminuses [1], which means that the entire population is well served [1]. Poorer inhabitants of the city can exchange recyclable materials or excess farm produce for tickets [1].

c) **Your answer:**
   - must include four or five highly-developed sentences
   - must name a global city (1 mark will be lost if no named example is given)
   - must explain how it is connected to the rest of the world, e.g. Airports allow global cities to be connected as they encourage the free movement of people between them and a sharing of cultures. Ports allow cities to be connected as they encourage trade between them, which encourages economic development.
   - might comment on the link between the city's global connections and its economy
   - must use geographical vocabulary accurately
   - must demonstrate good use of spelling, punctuation and grammar.

Question 4

a) The number of births per thousand [1] per year [1]

b) **Any three from:** Death rates; infant mortality rates; life expectancy; people per doctor; literacy rates; Human Development Index; access to clean water [3]

c) Gross domestic product (GDP) measures wealth alone [1] but does not describe how equally it is distributed [1]. GDP does not take into account whether the money created lots of jobs [1]. GDP is recorded for the population as a whole and hides possible gender inequality [1].

d) **Any two from:** Foreign investment; foreign government aid; foreign non-governmental organisation (NGO) aid; using intermediate technologies; fair trade; debt relief [2]

Question 5A

a) 400 m [1]

b) **Any three from:** Near to transport links; close to skilled workforce; cheaper land further away from CBD; brownfield land; government support / tax break; close to research intensive universities [3]

c) **Your answer must:**
   - include four or five highly-developed sentences
   - use geographical vocabulary accurately
   - include examples (1 mark will be lost if no relevant examples are given).

   Example: The growth of service industries, such as telephone banking in India [1], has meant higher paid jobs for the population, which has improved the standard of living [1]. More people working in higher paid jobs means that they can spend money elsewhere [1], which is known as the multiplier effect [1]. Improvements in education [1] mean that workers in India have more transferable skills [1]. Higher paid jobs mean more tax income for India as a whole [1], which can be spent on infrastructure, such as roads and sanitation [1].

d) **Any two from:** Recycling of heat from engines; use of wood from sustainable sources; building on brownfield land; paying living wage; generating own energy from renewables [2]

Question 5B

a) **Any one from:** Central Wales; west Wales; south-west Wales; north-east Scotland / rural Aberdeenshire; Western Isles of Scotland / Outer Hebrides; southern isles of Scotland / Argyll and Bute [1]

b) **Any three from:** Separation by sea; separation by highlands; proximity to urban areas; relative isolation from areas of higher population **[3]**

c) A = Total population, B = Death rate, C = Birth rate **[2]** **[1 mark for two correct]**

d) **Your answer:**
- must include four or five highly-developed sentences
- must suggest and explain strategies for dealing with the problems associated with an ageing population
- might consider the advantages and disadvantages of the strategies suggested.

Example: Increased maternity leave **[1]** could lead to an increase in the birth rate, as parents may decide to have more children / more people may decide to have children **[1]**. Increasing the retirement age **[1]** means that less money would need to be paid in pensions **[1]**. Training more specialist nurses and doctors to deal with elderly people **[1]** would ensure their needs are better met / take pressure off other areas of the health service **[1]**. Encouraging immigration of young adults **[1]** would lead to an increase in the number of economically active people **[1]**.

Question 6

a) Accept any location in Africa, India or south-east Asia **[1]**

b) **Any two from:** Birth rate; death rate; life expectancy; number of people per doctor; literacy rate; gender/sex ratio; percentage working in agriculture **[2]**

c) **Any four from:** Colonial control by other countries meant resources were removed **[1]**; after independence, debt was built up **[1]** trying to build infrastructure **[1]**; physical/natural barriers, such as deserts, adverse climatic conditions and rainforests, make development challenging **[1]**; unfair terms of trade **[1]**

d) **Any four from:** Wiping out debts owed by the developing country; issuing low-cost loans to the developing country; direct grants of money to the developing country; developed country builds infrastructure in developing country; instigating fair-trade schemes between countries **[4]**

e) **Any two from:** Once granted it is hard to track money; money can be lost to transport costs; corruption means money can be lost / used for other purposes; money may not go to those who need it most; people in donor countries may become frustrated **[2]**

f) **Your answer:**
- must include four or five highly-developed sentences
- must name a lower income (developing) country (1 mark will be lost if no named example is given)
- must give three or four different ways in which globalisation / international trade has contributed to development
- must explain in detail the effects of each one
- might give specific examples where appropriate
- might consider the costs and benefits of each one.

Example: In countries such as Uganda **[1]**, globalisation has meant that crops, such as coffee, can be sold globally **[1]**. With increased money coming into the country, Uganda has been able to build new road and rail links **[1]**. However, barriers to trade, such as tariffs, make some Ugandan products expensive, meaning there are fewer exports **[1]**. Cheaper shipping makes imports cheaper **[1]** so richer countries can dump overproduced crops in Uganda, putting local farmers out of work **[1]**. Cheaper iron and steel makes building infrastructure cheaper **[1]**. Fluctuating oil prices can mean cheaper or more expensive transport **[1]**.

Paper 3: Pages 215–238

Question 1A

a) B **[1]**; E **[1]**

b) **Your answer must:**
- demonstrate some data linkage, rather than lifting information straight from the source.

Any two from: All three rainforest-containing continents feature in the top 10; countries where the rate of loss has reduced are in South America; the trend, up or down, is not related to amount of clearance. **[2]**

c) **Your answer:**
- must consist of four or five well-developed sentences
- must make good use of the resources
- must give more than three valid reasons with explanations
- must use well-developed sentences
- must use relevant geographical vocabulary accurately
- might mention both plant and animal elements of diversity
- is likely to reference: low latitudes, size of area and features of the typical rainforest climate; relatively low population densities for a long period of time and the low level of technology associated with the indigenous populations having little impact.

d) Mean temperatures have increased over time / risen by about one degree since 1880 **[1]**; in an irregular pattern **[1]**

e) Accept answers between 1°C and 1.05°C **[1]** **[The units must be given to gain the mark. Do not accept answers of –0.42 to +0.62]**

f) 1930–1940 / 1930s **[1]**

Question 1B

a) 13 **[1]**

b) **Any two from:** Unevenly distributed **[1]**; more in the north and west / upland areas **[1]**; four in the south **[1]**

c) Lake District **[1]**

d) **Accept any valid reason and explanation, e.g.** Some national parks are nearer / better connected to larger urban areas **[1]**, where there will be more people available to easily visit **[1]**. Or Some are more famous due to particular points of interest / famous people, e.g. Snowdon is the highest mountain in Wales **[1]** so more people have heard of it and will want to visit **[1]**.

e) Depends on the ratio of number of visitors to permanent population / the Broads has a much smaller population than the number of visitors **[1]** so visitors will be more noticeable / have a bigger impact **[1]**.

f) Exmoor **[1]**

g) Point plotted at (9.5, 406) **[2]** **[1 mark for correct placement on the x-axis; 1 mark for correct placement on the y-axis]**

h) **Your answer must:**
- comment on the distribution of the points on the chart
- consider the reason behind this distribution.

Example: Generally, spending increases with visitor numbers / there is a positive correlation between spending and visitor numbers **[1]**. However, spending increases at a greater rate than numbers **[1]**, which could be because the bigger visitor numbers have caused more attractions/facilities/places to spend money to be built **[1]**.

Question 2A

a) Poorer countries suffer the most and richer countries suffer the least **[1]**, supported by data taken from the map and diagram (e.g. DR Congo – the poorest country in the chart at $311 GDP – will have wealth reduced severely, whereas the UK – shown richest on the chart – could experience an improvement in GDP) **[1]**.

Answers

b) Your answer must:
- consist of four or five well-developed sentences
- discuss a range of viewpoints, emphasising the conflicts/ differences
- discuss evidence from the resource sheet
- consider the long term versus the short term.

Possible approaches include:
- the suggestion that richer countries were just as responsible in the past for forest destruction as locals appear to be now (demand for forest products, etc.); also, emissions from long-industrialised countries have created problems in the rainforest and for the world, but have created high living standards – countries in Latin America, Asia and Africa might want a similar living standard / to catch up to some extent.
- the suggestion that richer countries have choices over employment / purchase of goods / way of life, which is not a luxury available to poorer people; the attraction of ranching, plantations, mining, logging, etc. is that it offers immediate reward, even though the forest suffers; it is difficult to take a long-term view when living in poverty; certified goods might carry a higher cost but do not necessarily create as many jobs as quickly.

c) Your answer:
- must relate the strategies to the resource material provided
- must clearly explain how the strategies apply to a number of issues at a range of scales: local/regional, national and global
- might include forest protection, preservation and regeneration, biodiversity, local economy, weather and climate (both organisations can contribute to each)
- must consider the weaknesses of each set of strategies
- must reach a conclusion about which organisation's strategies would be most effective
- demonstrate good use of spelling, punctuation and grammar.

Strengths: both sets of strategies will help with general rainforest issues, such as deforestation, impact on water supplies, weather and climate; they both work internationally so will raise awareness of the situations; their donations must largely come from richer countries so their advertising might cause people to change their ways of life.

Organisation A strengths: forest land is actually purchased so long-term control is possible; small parcels and new trees could fill in fragmented areas; focus on the forest gives prominence to biodiversity (reference to plants and animals); control is given to local groups who will understand the specific needs and way of life of native people; removed from government control so possibly less easily influenced; work with researchers to deepen knowledge and understanding; leading tourist walks to expand understanding to people outside the area.

Organisation B strengths: work with governments, so might be able to have greater power and influence and possibly avoid conflict; working with indigenous people maintains their identity; uses local knowledge (indigenous people); creates jobs and potential revenue/incomes for places with fragile economies; specific non-timber products reduce the threat to trees; marketing of products brings the rainforest issues to an international audience, which will increase knowledge and understanding beyond rainforest regions/ countries; modern technology is used effectively for the good of the people and territory.

Weaknesses: in both cases, the suggested weakness is likely to be about scale – the strategies can only affect a very small area; however, a large-scale impact is possible.

Organisation A weakness: lack of economic development.

Organisation B weakness: possible government manipulation.

Question 2B

a) Promoting the national parks will increase the number of visitors [1], which will make it much harder to conserve / keep safe the wildlife and landscape [1].

b) **Any three from:** Beautiful upland / mountain / lake views; it is possible to climb/sail/kayak, etc.; no traffic to pollute the air / cause noise; feeling of wilderness / open space / ability to be alone [**Accept any answer that places an emphasis on outdoors / fresh air / landscape and potential activities not available in urban areas**] [3]

c) The footpath has been widened by use [1] and has been covered/paved to strengthen it [1].

d) **Your answer must:**
- recognise that students can cause some physical damage, just like any other visitors, and that being in a large group can make students more intrusive than individuals
- mention that students are learning about the landscape, which should make them more appreciative of the fragility of the area
- comment on the economic contribution made by students, e.g. the contribution is not likely to be great but residential visits help to generate jobs in the centres
- consider the negative impacts of fieldwork activities, e.g. data collection activities can cause damage on river banks and slopes, surveys can cause disturbance to residents and other visitors, etc., and how they can be minimised through risk management
- demonstrate good use of spelling, punctuation and grammar.

e) **Any one from:** Water recycling and use of greywater in toilets; using locally produced goods in the café; selling clearly labelled local produce in the shop; encouraging the use of cycling and walking for access and public transport for longer journeys. [1]

Question 3

a) Measuring or counting a fraction / small part or group, which represents the whole place / population [1]

b) Taking information from along a line drawn on / from a map or along a road / other feature that is linear [1]

c) A random sample, e.g. asking every sixth person who passes irrespective of gender or age, would be appropriate because it is easily managed [1]. It is reliable because there is no bias or attempt to get the same number of people in different groups [1].

d) 25% [1]

e) A correctly proportioned area added to bar and key for Flat (subdivided house) [1]; a correctly proportioned area added to bar and key for Flat (purpose built) [1]

% Housing Type

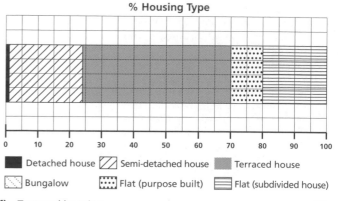

f) Terraced housing [1]

g) Trip line / Desire line [1]

h) **Any one from:** A larger town might have more choice; there might be a faster road; more frequent bus service; faster public transport **[T₁ is not the nearest town to village V, so any plausible reason to explain this different behaviour to the norm / expected from the map will be accepted]** **[1]**

i) Vertical axis correctly drawn and labelled **[1]**; horizontal axis correctly drawn and labelled **[1]**; suggestion of plotted points **[1]**

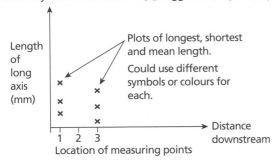

[Alternative answers such as a bar graph would also be acceptable]

j) B **[1]**

Question 4A

a) **Your answer:**
- must be primary data
- should be able to be reproduced from the description given
- should name and describe any equipment used
- could include a reference to accuracy and consistency.

b) **Your answer must:**
- be presentational, not statistical (e.g. pie chart, line graph)
- have processed information, not raw data
- include justification about: enabling trends or patterns to be identified easily; ability to use ICT; enabling information to be superimposed on to a base map.

c) **Your answer:**
- must be statistical (likely to be %, central tendency or dispersion or correlation (Spearman Rank))
- should quote data used to indicate appropriateness
- may mention result in justification.

d) **Your answer:**
- must state a specific location; named village, region of a named town
- should give two advantages which are specific to that place and not generic – these could be about the place (e.g. services all within a manageable area (described); CBD big enough for group work; sufficient variety of housing areas; or about their own location (journey time, etc.)

e) **Any risk relating to the answer given for part d) (could be for any element of the work), e.g.** traffic or abuse from members of the public if carrying out a questionnaire. **[1]**

f) **Two valid points relating directly to the answer given for part e) (should not simply repeat the description), e.g.** for traffic: using zebra crossings if available and not attempting to cross near busy corners. Or for abuse from members of the public: wearing school uniforms, working in pairs and carrying school identification. **[2]**

g) **Your answer:**
- must include four well-developed sentences
- must clearly state the conclusions reached (conclusions need to be specific rather than general)
- should relate each conclusion to the result
- should relate each conclusion to the geographical theories/concepts underpinning the enquiry.

Question 4B

a) One size might be too small or too big for the conditions that day **[1]**. It makes it possible to test the idea that smaller pebbles will travel further than larger ones **[1]**.

b) Groynes trap material and interfere with longshore drift **[1]** so the results would not represent the uninterrupted work of the sea **[1]**.

c) **Any two from:** Some may be nearer or further away than were noticed; some could have moved in the opposite direction to the majority; some could be lost in the water or under piles of other pebbles; students may not have seen them / lack of observation **[2]**

d) Any diagram or sketch which shows the different size classes of the pebbles (this could be using symbols or colours) **[1]**; origin clearly indicated **[1]**; horizontal axis clearly labelled **[1]**; vertical axis clearly labelled **[1]** **[Credit can also be given for a key]**

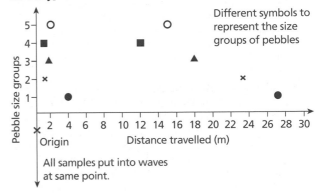

e) **Your answer:**
- must show an understanding of how each average is calculated
- must discuss how appropriate each average is for the data
- might consider how each average would be used in further calculations / presented to show useful results, e.g. the mean can be used for further statistical tests.

Example: The mean would be influenced by extreme values **[1]**, e.g. if only one pebble in a size group travelled much further than the rest **[1]**. There is not a lot of data, e.g. in size group 2, so there may not be a single modal distance **[1]**. Distance categories might have to be made too big to be useful **[1]**. The median is not affected by extreme values, so would be useful if there was one pebble in a group that had gone much further than the rest **[1]**, but the median cannot be used for any further calculations that students want to do **[1]**.

f) **Any four from:** Checking tide times; looking for rips; studying wave behaviour first to check likelihood of water rushing on to students without warning; not having lone students at any point in the measuring zone; ensuring no one is carrying unnecessary bulk / weighty/unbalanced bags; wearing appropriate footwear and clothing to avoid tripping or overbalancing; having 'look-out' students when others are throwing in or retrieving pebbles, so that they can warn of unusual/changing waves; measuring longshore distance away from the water's edge, so that students are not inundated. **[4]**

Question 4C

a) To ensure consistency and reliability / There are a number of people collecting data to be used together, so they must all do the same thing every time **[1]**

b) For safety **[1]**. River conditions can change quickly and unexpectedly; watching the water approach means any changes in speed / debris which could hit someone will be seen **[1]**.

Answers

c) Regular intervals: easy and quick to mark out places across the channel / the whole channel width can be covered **[1]**; very few measurements will be taken at narrow sites / could be a very big difference in the amount of data collected at different sites **[1]**

Same number at each site: amount of data at each site will be consistent / the whole channel width can be covered **[1]**; the width of the channel needs to be divided and the interval calculated – could therefore take longer and have greater margin for error / at narrow sites the places could be very close together, so if bedload is big, the same piece could be at more than one measuring point **[1]**

d) **Your answer must:**
   • consider physical issues and technical problems
   • explain each problem in detail.

   Example: Identifying which sample should be measured can be difficult **[1]**, especially if the bedload is small and there are a number of particles directly below the point chosen **[1]**. Large bedload cannot be easily/safely removed from the channel for accurate measurement **[1]**. In faster flow, students may struggle to stand still **[1]** and not disturb the bedload **[1]**. Very small particles are difficult to accurately measure **[1]**.

e) It is qualitative rather than quantitative / It depends on people's judgement (subjective) rather than a measurement (objective).   **[1]**

f) Any diagram that clearly shows which site the data is from and the difference between the three measurements, with a clearly labelled horizontal axis **[1]**; clearly labelled vertical axis **[1]**; accurate plots, which could be coloured or given different symbols **[2]**

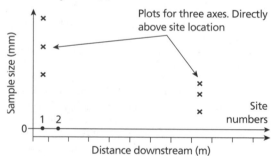

g) **Accept any sensible suggestion for a test to see if the increased discharge from a single storm will affect the river's ability to move material, e.g.** Pebbles in a range of size classes could be painted and put into the river at the same safe point **[1]**. After the storm / the next day, the distance any of the pebbles have been carried can be measured **[1]**.